ANALYTICAL CHEMISTRY SYMPOSIA SERIES — volume 8

ion-selective electrodes, 3

*Third Symposium held at Mátrafüred, Hungary,
13—15 October, 1980*

ANALYTICAL CHEMISTRY SYMPOSIA SERIES

Volume 1   Recent Developments in Chromatography and Electrophoresis. Proceedings of the 9th International Symposium on Chromatography and Electrophoresis, Riva del Garda, May 15-17, 1978
edited by A. Frigerio and L. Renoz

Volume 2   Electroanalysis in Hygiene, Environmental, Clinical and Pharmaceutical Chemistry, Proceedings of a Conference, organised by the Electroanalytical Group of the Chemical Society, London, held at Chelsea College, University of London, April 17-20, 1979
edited by W. F. Smyth

Volume 3   Recent Developments in Chromatography and Electrophoresis, 10. Proceedings of the 10th International Symposium on Chromatography and Electrophoresis, Venice, June 19-20, 1979
edited by A. Frigerio and M. McCamish

Volume 4   Recent Developments in Mass Spectrometry in Biochemistry and Medicine, 6. Proceedings of the 6th International Symposium on Mass Spectrometry in Biochemistry and Medicine, Venice, June 21-22, 1979
edited by A. Frigerio and M. McCamish

Volume 5   Biochemical and Biological Applications of Isotachophoresis. Proceedings of the First International Symposium, Baconfoy, May 4-5, 1979
edited by A. Adam and C. Schots

Volume 6   Analytical Isotachophoresis. Proceedings of the 2nd International Symposium on Isotachophoresis. Eindhoven, September 9-11, 1980
edited by F. M. Everaerts

Volume 7   Recent Developments in Mass Spectrometry in Biochemistry, Medicine and Environmental Research, 7. Proceedings of the 7th International Symposium on Mass Spectrometry in Biochemistry, Medicine and Environmental Research, Milan, June 16-18, 1980
edited by A. Frigerio

Volume 8   Ion-Selective Electrodes. Third Symposium held at Mátrafüred, Hungary, 13-15 October, 1980
edited by E. Pungor and I. Buzás

ANALYTICAL CHEMISTRY SYMPOSIA SERIES — volume 8

ion-selective electrodes, 3

Third Symposium held at Mátrafüred, Hungary,
13—15 October, 1980

editor
Prof. E. Pungor, Ph. D., D. Sc.
Member of the Hungarian Academy of Sciences

associate editor
I. Buzás, Ph. D., C. Sc.

ELSEVIER SCIENTIFIC PUBLISHING COMPANY
Amsterdam – Oxford – New York   1981

ORGANIZING COMMITTEE OF THE SYMPOSIUM

President: Prof. E. PUNGOR, Ph. D., D. Sc., Member of the Hungarian Academy of Sciences;
Secretary: K. TÓTH, Ph. D., C. Sc., GY. FARSANG, Ph. D., C. Sc.; *Members:* ZS. FEHÉR, Ph. D., C. Sc.,
G. NAGY, Ph. D., C. Sc.; *Organizer:* E. GRÁF, Ph. D.

Joint edition published by

Elsevier Scientific Publishing Company, Amsterdam, The Netherlands and

Akadémiai Kiadó, The Publishing House of the Hungarian Academy of Sciences, Budapest, Hungary

The distribution of this book is being handled by the following publishers
for the U.S.A. and Canada
Elsevier/North-Holland, Inc.
52 Vanderbilt Avenue
New York, New York 10017, U.S.A.

for the East European Countries, Democratic People's Republic of Korea, People's Republic of
China, People's Republic of Mongolia, Republic of Cuba and Socialist Republic of. Vietnam
Akadémiai Kiadó, The Publishing House of the
Hungarian Academy of Sciences, Budapest

for all remaining areas
Elsevier Scientific Publishing Company
P.O. Box 211, 1000 AE Amsterdam, The Netherlands

Library of Congress Cataloging in Publication Data
Main entry under title:

Ion-selective electrodes.

    (Analytical chemistry symposia series; v. 8)
    Bibliography: p.
    Includes index.
    1. Electrodes, Ion selective—Congresses. I. Pungor,
E. (Ernő) II. Buzás, I. III. Series.
QD571.I593   541.3'724   81-15179
ISBN 0-444-99714-8 (U.S.)   AACR2

ISBN 0-444-99714-8 (Vol. 8)
ISBN 0-444-41786-9 (Series)

CONTENTS

Plenary Lectures

Keynote Lectures

Panel Discussion

PREFACE

For the third time, a Symposium on Ion-Selective Electrodes was organized at Mátrafüred by the Analytical Chemical Committee of the Hungarian Academy of Sciences.

The form of the Symposium, adopted ever since the first of these meetings, again proved to be very effective. The method employed implies that the lectures only serve as starting points for the discussion rather than being a simple statement of final results, as is common at many conferences. The time spent on presentation and on discussion was roughly equally divided between the two.

As with the previous symposia, a special topic was selected with which to end. This was then discussed by the participants with the intention of making it a principal object of further research. One of the most exciting current problems is that of the standardizing of ion-selective electrodes, and this topic was introduced by an internationally recognized expert, Professor Bates, whose presentation elicited a lively response.

As before, the material presented at the Symposium is here published by the Publishing House of the Hungarian Academy of Sciences jointly with Elsevier Scientific Publishing Company. It is our fervent hope that this publication of the lectures and discussions will be found valuable by all those engaged in research activity in the field, and those interested in the analytical applications of ion-selective electrodes.

We hope that it will help to stimulate even greater interest in ion-selective electrode development and application.

E. Pungor

LIST OF PARTICIPANTS

Ammann, D. (Switzerland)
Bálint, T. (Hungary)
Bartalits, L. (Mrs) (Hungary)
Bates, R. G. (USA)
Bertényi, I. (Hungary)
Boksay, Z. (Hungary)
Boran, R. B. (GDR)
Bouquet, G. (Hungary)
Buck, R. P. (USA)
Burger, K. (Hungary)
Deák, É. (Mrs) (Hungary)
Domokos, L. (Hungary)
Ebel, M. F. (Mrs) (Austria)
Fehér, Zs. (Mrs) (Hungary)
Fjeldly, T. A. (Norway)
Gábor, T. (Mrs) (Hungary)
Garai, T. (Hungary)
Gracza, M. (Mrs) (Hungary)
Gráf, Z. (Mrs) (Hungary)
Gratzl, M. (Hungary)
Gyenge, R. (Mrs) (Hungary)
Hankó, K. (Mrs) (Hungary)
Hopkala, H. (Mrs) (Poland)
Hulanicki, A. (Poland)
Jänchen, M. (GDR)
Johansson, G. (Sweden)
Johansson, K. (Finland)
Jovanovic, M. S. (Yugoslavia)
Józan, M. (Hungary)
Juhász, E. (Hungary)
Koryta, J. (Czechoslovakia)
Lewenstam, A. (Poland)
Lindner, E. (Hungary)

Mezei, P. (Hungary)
Morf, W. E. (Switzerland)
Müller, H. (GDR)
Nagy, G. (Hungary)
Nagy, K. (Norway)
Pál, F. (Hungary)
Pethő, G. (Hungary)
Petr, J. (Czechoslovakia)
Petrukhin, O. M. (USSR)
Porjesz, E. (Mrs) (Hungary)
Pretsch, E. (Switzerland)
Pólos, L. (Hungary)
Pungor, E. (Hungary)
Radovanovicz, M. (Mrs) (Yugoslavia)
Rakiás, F. (Hungary)
Scherr, Z. (Mrs) (Hungary)
Senkyr, J. (Czechoslovakia)
Siemroth, J. (GDR)
Simon, W. (Switzerland)
Szepesváry, P. (Mrs) (Hungary)
Szűcs, Z. (Mrs) (Hungary)
Tenno, T. (USSR)
Thomas, J. D. R. (Wales)
Tomcsányi, L. (Hungary)
Tóth, K. (Mrs) (Hungary)
Trojanowicz, M. (Poland)
Umezawa, Y. (Japan)
Varga, M. (Hungary)
Vasilikiotis, G. (Greece)
Vesely, I. (Czechoslovakia)
Virtanen, R. (Finland)
Vlasov, Yu. G. (USSR)
Werner, G. (GDR)

PLENARY LECTURES

THERMODYNAMIC BEHAVIOR OF ION-SELECTIVE ELECTRODES

ROGER G. BATES

Department of Chemistry, University of Florida, Gainesville, Florida 32611, USA

ABSTRACT

Although ion-selective electrodes (ISE) have provided a valuable tool for the determination of ion concentrations, these sensors respond most directly to some function of ion activities. As such, they have great possibilities, as yet largely unexplored, in the determination of precise thermodynamic quantities for ionic systems. As a first step, one must evaluate the quality of the electrode response; this is often difficult to accomplish, for scales of single ion activity are not uniquely defined by thermodynamics. If the electrode is combined with a second electrode of demonstrated reliability, however, thermodynamic data for neutral ion combinations can be obtained. When these data agree with accepted values resulting from other precise thermodynamic measurements, one has evidence that the ISE is functioning in a reversible manner.

This procedure for characterizing the thermodynamic behavior of ion-selective electrodes will be illustrated and examples given of data for systems difficult to study by conventional techniques. An attempt is also made to evaluate the extent to which certain electrodes function thermodynamically over ranges of temperature and in nonaqueous and partially aqueous solvents. A novel and convenient relative method, making use of the Na-ISE, for determining the dissociation constants of weak bases in water, water/methanol mixtures, and deuterium oxide over the range 15 to $40^{\circ}C$, is described.

3

INTRODUCTION

The development of electrochemical sensors selective for
one species of ion or even for a variety of specific compounds
of critical importance in biomedicine and industrial processes
has proved one of the exciting events of the past 15 years.
Analytical chemists in particular have been able to profit
from the availability of ion-selective electrodes which offer
a convenient and rapid means of estimating the concentrations
of certain elements which formerly were difficult to determine.
The practical results have been enormously beneficial to the
laboratory scientist. In addition, the development of the
theory of operation of these sensors has led to an increased
understanding of membrane behavior in general and of the mechanisms of transfer through membranes important to human physiology (1,2). Extensive and detailed studies of the selectivity of these electrodes (3) add much to this understanding
and aid materially in increasing the accuracy of analytical
determinations of ion concentrations.

When the transfer process is viewed from the macroscopic
vantage point, it becomes clear that the electrochemical response of a membrane electrode is a function of the activities
of the selected ion on the two sides of the membrane and of
the rates of mass transfer through the membrane (4). In other
words, the electrode is a device permitting the transport of
a single species of ion between two states of differing Gibbs
energy. This restriction has both advantages and disadvantages.
If the analytical chemist wishes to determine ion concentrations in one of the separated phases most directly, some pro-
vision for equalizing activity coefficients must be made.
However, these sensors also provide the possibility of determining very conveniently the activity function and other properties of systems related to the free energy.

In the application of ion-selective electrodes to determine the thermodynamic properties of solutions, no new techniques are usually involved. Ion-selective electrodes may,
however, be extremely useful as replacements for more conventional electrodes that have found extensive use for many years.

4

For example, Harned and his co-workers (5) led the way in
applying the cell with hydrogen electrodes and silver-silver
halide electrodes

$$Pt;H_2(g,1 \text{ atm}) | \text{Soln.},H^+ \text{ and } X^- | AgX;Ag \qquad (A)$$

to the determination of activity coefficients for the halogen
acids (HX) in water and water/organic solvents alone and in
mixtures with other electrolytes, as well as for the study of
weak acid equilibria and associated thermodynamic properties
in a variety of solvent media.

In spite of its versatility, this cell has its limitations.
The hydrogen electrode is subject to poisoning by heavy metals
and certain other materials, while some organic acids are reduced at the platinum surface. The proper functioning of silver halide electrodes requires the establishment of a simple
solubility equilibrium governed by the solubility product constant. This equilibrium is disturbed by ammonia and amines
and by anions (such as sulfide) which form silver compounds of
lower solubility than the halide. In some aprotic media,
chloro complexes of high stability may be formed (6). These
difficulties make it impossible to determine accurate thermodynamic data with the hydrogen-silver halide cell in every instance, and then ISE's may offer a successful substitute.

Inasmuch as the mechanism of operation of a membrane electrode is one of ion transfer, these electrodes may be useful
in oxidizing and reducing media which would affect adversely
those electrodes whose potentials depend on electron transfer.
Nevertheless, the standard potential of a given type of membrane electrode is not usually fixed and may fluctuate to some
extent with time. Furthermore, the change of electrode potential with changes in the logarithm of the activity of the selected ion may not conform to theoretical expectations and thus
may require calibration. In this regard, the extensive program of Pungor and his associates in the Budapest laboratory
for determining and collecting selectivity coefficients (7) is
of the highest usefulness. For these reasons, the use of ISE's
to determine reliable thermodynamic data should be preceded by
an examination of the behavior of the cell system involved.

It is the purpose of this paper to review some of the most
important thermodynamic applications of ion-selective electrodes and to outline studies that enable one to establish
the suitability of these sensors for thermodynamic measurements, together with the limits of their applicability. The
status of this subject was capably summarized by Butler in
1969 (8). Many more recent studies provide valuable guidance
bearing on this problem, and it is not possible here to give
an exhaustive review of all that has been done. Instead,
these procedures will be illustrated by selected examples.

FREE ENERGY OF AQUEOUS SYSTEMS

1. Activity Coefficients

A comparison of activity coefficients determined with the
aid of ion-selective electrodes and those based on emf data
of proved reliability or on static or isopiestic vapor pressure methods provides one of the most direct ways of assessing
the performance of the ISE. To avoid the uncertainties of
liquid junctions, cells without transference should be chosen.
The response of the pH glass electrode, the earliest of the
current group of ISE's, can, for example, be readily examined
by measurement of the emf of the cell

$$\text{pH glass} | \text{HCl}(m) | \text{AgCl;Ag} \tag{B}$$

over wide ranges of the molality m. The AgCl;Ag electrode is
known to display a Nernstian response up to m=3 or higher.
The mean ionic activity coefficient ($\gamma_\pm$) of HCl is given by

$$-\log \gamma_\pm = (E-E^o)/2k + \log m \tag{1}$$

where k is written for the Nernst slope $(RT \ln 10)/F$. Inasmuch as E^o (the standard potential) for the ISE is unknown, it
must be determined by a suitable extrapolation to m=0 or calculated from measurements of E at a low molality where $\gamma_\pm$ is
precisely known (9). Since these values of the activity coefficient were derived from measurements of cell A, it is sufficient to show that E_A-E_B is constant, independent of m.
Oddly, neither type of measurement is commonly used to check
the response of pH glass electrodes, for which purpose pH

buffers are usually chosen, in cells with liquid junction.

On the other hand, the sodium glass electrode has been employed rather extensively to determine activity coefficients of sodium salts in aqueous solutions (10-13). By use of the cell

$$Na-ISE|Soln.,Na^+ \text{ and } Cl^-|AgCl;Ag \qquad (C)$$

the Na-ISE was shown to yield activity coefficients for NaCl in agreement with accepted values derived by other thermodynamic methods. This agreement is apparent in Table 1, where the results of Schindler and Wälti (13) are compared with the activity coefficients of NaCl determined by isopiestic vapor pressure methods (14). The Na-ISE was standardized in a relatively concentrated solution (molality 0.1007 mol kg^{-1}), for which the activity coefficients from vapor pressure measurements were adopted.

Data for the activity coefficient of calcium chloride in its aqueous solutions (triangles) derived from a cell with the Orion 92-20 calcium electrode combined with the AgCl;Ag electrode (15) are shown in Figure 1. The solid line is drawn through the accepted values over the experimental range, and the reference point is the solution of I=0.01. Similar measurements can be used to confirm simultaneously the thermodynamic behavior of two ion-selective electrodes, as also shown in Figure 1 (15), based on the cell

$$Na-ISE|NaF(m)|F-ISE \qquad (D)$$

Cell D was standardized in a 0.01m solution of NaF. Again, the reasonable agreement between the experimental points (circles) and accepted data (solid line) shows that these electrodes are functioning thermodynamically in the range 0.001<I<1 mol kg^{-1}.

Figure 2 shows a comparison of the activity coefficients of choline chloride (trimethyl-β-hydroxyethylammonium chloride) derived from emf measurements of the cell

$$Choline-ISE|Choline chloride(m)|AgCl;Ag \qquad (E)$$

with those determined by the isopiestic method over a wide range of molality (16). It appears that this electrode also

2*

functions reversibly in choline chloride solutions.

Inasmuch as activity coefficients at 25°C of dozens of electrolytes in aqueous solutions of concentration exceeding 0.1m have been determined precisely (14), this procedure may have wide application in testing the behavior of other electrodes. For acceptable results, it is necessary that no side reaction with the second (reference) electrode occur. Unfortunately, a lack of extensive data for activity coefficients precludes the application of this procedure to many nonaqueous and mixed solvents and at temperatures other than 25°C.

2. Mixtures of Electrolytes

Many electrodes are remarkably selective for one species of ion, yet for accurate measurements the need for selectivity coefficients is general. Furthermore, interest in the specific interactions of ions in salt mixtures has been increasing steadily in recent years, and notable progress has been made in accounting for activity-coefficient behavior in electrolyte mixtures (17-19). Ion-selective electrodes offer a means of providing the data needed to enlarge the scope of these studies. At the same time, data already available can be used to confirm the response of an ISE in the presence of a possible interferent. This procedure is, however, of limited applicability at the present time, but an example will illustrate its nature.

Activity coefficients of acids HX in the presence of a number of other electrolytes, especially salts MX_n, have been determined by emf methods. In general, cell A was used. Substitution of the pH glass electrode for the hydrogen electrode, as in cell B, would evidently enable one to compare the behavior of the glass electrode with the hydrogen electrode in a mixture of electrolytes. A more interesting application would utilize the activity coefficients of NaCl and $CaCl_2$ in their mixtures, determined by isopiestic vapor pressure measurements (20) or with sodium amalgam electrodes (21). Figure 3 shows the variation of log γ_{NaCl} and log γ_{CaCl_2} at 25°C with the composition of the mixture at a total fixed ionic strength (I) of 3.0 mol kg^{-1}. The linear Harned rule is obeyed for both components at this value of I:

$$\log \gamma_{NaCl} = -0.1463 + 0.0018I_{CaCl_2} \qquad (2)$$

$$\log \gamma_{CaCl_2} = -0.3010 + 0.0165I_{NaCl} \qquad (3)$$

where I_{NaCl} and I_{CaCl_2} are the contributions of the two salts to the total ionic strength (3.0) and the first terms on the right are the values of $\log \gamma_{NaCl}$ and $\log \gamma_{CaCl_2}$ in pure solutions of these salts at $I=3.0$. To check the performance of the Ca-ISE in the presence of NaCl, one prepares the following cell

$$\text{Ca-ISE} | \text{NaCl}(m_1), \text{CaCl}_2(m_2) | \text{AgCl;Ag} \qquad (F)$$

containing a series of solutions of constant total ionic strength, $I=3.0=m_1+3m_2$. The mean activity coefficient of $CaCl_2$ is given by

$$-\log \gamma_{CaCl_2} = \frac{2}{3}\frac{E-E^{\circ}}{k} + \frac{1}{3}\log\left[m_2(m_1+2m_2)^2\right] \qquad (4)$$

and E° is readily obtained by measuring E in a pure $CaCl_2$ solution ($m_2=1.0$, $m_1=0$, $I=3.0$) for which γ_{CaCl_2} is known from Equation 3.

Although it is often not possible to make a thorough-going confirmation of thermodynamic behavior in advance, ion-selective electrodes have proved useful in the study of salt mixtures. One may mention the use of the Na-ISE to study the properties of mixtures of NaCl and tris(hydroxymethyl)aminomethane hydrochloride (Tris·HCl) (13) and the study of mixtures of NaCl with eight different salts of the 1:1, 1:2, and 2:1 valence types by Lanier (22). In these instances, the adherence to Harned's rule offers some confirmation of the proper functioning of the electrode.

9

1. Activity Coefficients

Ion-selective electrodes offer promise in the study of the Gibbs energies of compounds and processes in certain nonaqueous media and in mixed solvent systems, especially those containing water as one component. This type of application may not be successful, however, if the electrode is of the liquid ion-exchanger type, where the membrane contains a nonaqueous liquid which must remain insoluble in the test solutions. Thus, "solid-state" electrodes may be better suited to such use.

Unfortunately, the isopiestic vapor pressure method has never been adapted successfully to measurements of free energies in nonaqueous media, and only a few precise emf measurements have been made in these solvents with amalgam electrodes. Consequently, it is difficult to calibrate the behavior of ISE's by a comparison of activity-coefficient data as one can do in the aqueous medium.

Nevertheless, useful information has been obtained in alcohol/water solvents, especially with the fluoride (LaF_3) electrode, the sodium glass electrode, and the cationic glass electrode. The activity coefficients of NaCl were measured with the cationic electrode in water/methanol and found to be in agreement with data obtained with amalgam electrodes (10,23-25). The Nernstian response of the F-ISE has also been verified in 60% ethanol (26). The lithium response of the Beckman cation-sensitive glass electrode has been studied by Mukherjee in 50% ethanol (27) and in propylene carbonate (28) and the effects of interfering ions analyzed in terms of an ion-exchange model. In addition, Nakamura (29) has shown that the response of this electrode to Li, Na, K, Cs, Tl, and Ag ions parallels that of amalgam electrodes in seven media including the donor solvents DMF and DMSO as well as the aprotic solvents acetonitrile and propylene carbonate.

Lanier's work (24) provides an illustration of procedures that can be recommended. Cell C, free from the uncertainties of the diffusion potential, was used successfully in several water/organic solvent mixtures. Since the AgCl;Ag electrode is known to function reversibly in many mixed solvents, the activity coefficient of NaCl in these media is given by

$$-\log \gamma_{NaCl} = (E-E^o/2k + \log m \qquad (5)$$

where m is the molality of NaCl. Both the Na-ISE and the cationic glass electrode were used. The difficulty arises in determining the standard emf E^o for the cell, in the absence of independent information for γ_{NaCl} in a reference solution. Lanier chose to extrapolate values of log γ^* (referred to E^o based on γ_{NaCl} in the aqueous medium) to m=0, thus evaluating the medium effect.

Recently Yang et al (30) have reported the use of a Na-ISE and the cationic electrode to determine the activity coefficients of NaCl and KCl in water/ethanol mixtures from 10 to 90 wt. % ethanol. Lacking values for reference solutions in the mixed solvents, they based their values in all solvents on an aqueous scale. Transfer activity coefficients are needed in order to relate their values to infinite dilution in the water/ethanol solvents; nevertheless, they were able to show that their results were consistent with the vapor pressures of the solutions, thus confirming the behavior of their electrode systems.

2. Transfer Free Energies

Numerical values of the activity of an electrolyte i consisting of ν ions are fixed by the arbitrary choice of a reference state in which the partial molal Gibbs energy is designated μ^o:

$$\mu_i = \mu_i^o + \nu RT \ln a_i \qquad (6)$$

In this standard state the activity a_i is unity. In view of the desirability that activity shall approach the molality in dilute solutions, it is customary to use different standard states for each solvent medium. For a transfer process

i (std. state, solv. 1) = i (std. state, solv. 2)

the transfer Gibbs energy ΔG_t^o is

$$\Delta G_t^o = \mu_2^o - \mu_1^o = \nu RT \ln \gamma_t \qquad (7)$$

where γ_t is the "transfer activity coefficient". For a cell such as cell C, the transfer from water (w) to another solvent (s) is given by

$$\Delta G_t^O = \mu_s^O - \mu_w^O = F(E_w^O - E_s^O) \qquad (8)$$

The standard emf (E^O) is often determined from measurements of emf in dilute solutions by extrapolation to I=0, aided by activity coefficients estimated by the Debye-Hückel theory.

The transfer energy can, in principle, also be derived from solubilities in the two media (w and s) of the electrolyte to which the cell is reversible, and from vapor pressures. Thus, the measurement of ΔG_t^O is a possible means of assessing the response of an ISE in a nonaqueous or mixed solvent.

Transfer energies of the alkali chlorides have been determined from emf measurements with amalgam electrodes (31-33), by measurements of vapor pressure (34), and from the potentials of cationic glass electrodes (35-37). Table 2 compares the values obtained by ISE measurements with other literature data.

Most of these studies of the behavior of ISE's in nonaqueous solvents have dealt with salts of univalent cations. In a recent noteworthy investigation, Coetzee and Istone (40) have used the Cu-ISE to evaluate Gibbs energies of transfer for copper ion from water to other solvents. Although this procedure is essentially non-thermodynamic, the results were eminently reasonable and consistent with the numbers derived on the basis of the equivalence of the transfer energies of tetraphenylarsonium and tetraphenylborate ions, an assumption that is finding increasing support (41).

EQUILIBRIUM DATA

Ion-selective electrodes also show promise of many useful applications in the determination of equilibrium constants. A wealth of accurate data for equilibrium systems already exists, and consequently the thermodynamic behavior of the ISE can often be evaluated by a comparison with accepted results. Many weak acid-base systems have been studied, for example,

with cell A, utilizing a hydrogen gas electrode. The glass
electrode, although less precise, is much more versatile than
the hydrogen electrode. Measurements of cell B can therefore
provide information for systems whose redox properties preclude
the use of the $Pt;H_2,H^+$ couple.

The pK of an acid-base dissociation equilibrium HA,A is
given in terms of the emf of either cell A or cell B by

$$pK = \frac{E-E^o}{k} + \log m_{Cl} + \log \frac{m_{HA}}{m_A} + \log \frac{\gamma_{HA}\gamma_{Cl}}{\gamma_A} \qquad (9)$$

when the cell solutions contain HA, A, and Cl^-. If the glass
electrode is used, the standard emf E^o is unknown and variable;
consequently, the cell must be standardized frequently, either
by measurement of E in dilute HCl solutions of known γ_{HCl}
(compare Equation 1) or in a weak acid system of known pK.

Electrodes responding to the anion of a weak acid are often
particularly useful in determining the pK for that acid-base
pair. Thus, the LaF_3 electrode is well suited to the study of
fluoride complexes with H^+ (5,42), and the Ag_2S electrode for
similar studies in sulfide systems.

Ion-selective electrodes respond to the activity of free
ions and hence are suitable for the study of ion pairing or
complexation equilibria. Rechnitz and his co-workers, for
example, have used the Ca-ISE to investigate the stability of
a variety of calcium complexes (43,44), and electrodes selective for Cu, Pb, and Cd suggest themselves for similar purposes.

Although the AgCl;Ag electrode or a Cl-ISE is extremely
useful in the determination of pK values for weak acids of the
charge type HA^o,A^- by means of the Harned cell A or cell B,
this method is less advantageous for weak base systems, charge
type HA^+,A^o. The reason is clearly evident from Equation 9:
for the former (acetic acid type), the last term of Equation 9
is small and varies linearly with ionic strength; for the latter (ammonium type), a considerable variation in the last term
and in pK with ionic strength is inevitable. Furthermore,
complexation of AgCl with many weak base donors constitutes an
experimental problem. As a result, the thermodynamics of weak

base equilibria has been accorded less attention than that of weak acids.

Some years ago it was recognized that the sodium amalgam electrode would be a suitable replacement for the AgCl;Ag electrode in cell A for determining the strengths of weak bases. Recent studies by K. Tanaka in the author's laboratory (45) have shown that the Na-ISE can profitably be used for this purpose. It will be remembered that the thermodynamic pK value of a protonated base BH^+ is determined by extrapolation to I=0. If cell A or cell B is used, the emf for a constant chloride concentration and buffer ratio changes considerably with I, as shown in the upper curve of Figure 4. However, as the bottom curve of the figure shows, the emf of the cell

$$\text{pH glass} | BH^+, B, Na^+ (0.01m) | Na\text{-ISE} \qquad (G)$$

changes but slightly with I and in linear fashion. The pK_a of BH^+ is related to the emf of cell G by

$$pK_a = \frac{E-E^O}{k} - \log m_{Na} + \log \frac{m_{BH}}{m_B} - \log \frac{\gamma_B \gamma_{Na}}{\gamma_{BH}} \qquad (10)$$

In this instance, the activity coefficient term is small, as a result of the approximate equivalence of γ_{Na} and γ_{BH}.

The most effective use of cell G involves the use of a reference base of known pK_a, in lieu of other methods of obtaining the standard emf E^O. For this purpose, Tris, whose pK_a in water (46,47) and in 50% methanol (48) has been determined over a range of temperatures, was chosen. The molality of Na^+ was 0.01 mol kg^{-1} in all solutions, and the buffer ratio was always unity, subject only to small corrections for solvolysis. Thus, if R refers to the reference base,

$$pK_a = pK_R + \left(\frac{E-E_R}{k}\right)_{I=0} \qquad (11)$$

where the difference in emf is obtained from the intercepts of E vs. I plots at I=0. Typical curves for the two bases Tris (R) and 2-aminopyridine (2AmPy) are shown in Figure 5. Straight lines of small slope were usual.

Preliminary results of applying this convenient and rapid method to the determination of the pK_a values for several protonated bases whose thermodynamic properties have been determined from earlier measurements with the hydrogen electrode are given in Table 3. The acceptable agreement between the values obtained with cell G and by other thermodynamic procedures testifies to the validity of the new method. It also shows that these ISE's behave satisfactorily at temperatures from 15 to 35°C. Similar data for "Bis" (2-amino-2-methylpropanediol), the bis(hydroxymethyl) analog of Tris, in Table 4 show that these electrodes have a theoretical response in 50% water/methanol solvents as well.

The pH glass electrode is known to respond to D^+ in D_2O (55-57). Unfortunately, the pK_a of Tris in D_2O has not been determined with the deuterium gas electrode. Nevertheless, the pK values of primary phosphate ion in D_2O have been measured (58), and they permit the standard emf (E_B^O) of cell B to be determined, with the aid of which the pK_a for Tris has now been obtained. The results are shown in Table 5; they lead to an enthalpy of dissociation of 11,920 cal mol^{-1} at 25°C. A recent calorimetric determination of the heat of neutralization of Tris in D_2O (59) yielded a value of 11,640 cal mol^{-1}.

The behavior of the Na-ISE in D_2O solutions of Tris buffers can be examined by determining pK_a from the emf of cell G, containing Tris·H^+, Tris, and Na^+, molality 0.01 mol kg^{-1}. The calculation is made by Equation 10, but in this instance the standard emf (E_G^O) must be obtained by independent means, as follows. The standard emf (E_C^O) of cell C containing a dilute (0.01m) solution of NaCl was calculated from a knowledge of the activity coefficient of NaCl. At an ionic strength of 0.01, the latter can be obtained with little uncertainty from the Debye-Hückel formula. With a knowledge of E_B^O from a measurement of cell B with phosphate buffers in D_2O, one can derive E_G^O:

$$E_G^O = E_B^O - E_C^O \tag{12}$$

and from it pK_a by extrapolation to I=0. The result of this second determination of the pK_a of Tris in D_2O at 25°C is also given in Table 5. It is apparent that the response of the Na-ISE in D_2O at 25^OC is confirmed.

The examples cited here serve to emphasize that ISE's have a vast potential for the acquisition of useful thermo-dynamic data. For the greatest reliability, however, it is essential that the quality of the electrode response be carefully examined, in order to justify confidence in the results obtained.

ACKNOWLEDGMENT

This work was supported in part by the National Science Foundation (U.S.A.) under Grant No. INT78 11287.

REFERENCES

1. G. Eisenman, G. Szabo, S. Ciani, S. McLaughlin, and S. Krasne, in Progress in Surface and Membrane Science, Vol.6, (J. F. Danielli, M. D. Rosenberg, and D. A. Cadenhead,eds.). Academic Press, New York, 1973, p. 139.
2. R. P. Buck, Crit. Rev. Anal. Chem. 5, 323 (1975).
3. E. Pungor and K. Tóth, Anal. Chim. Acta 47, 291 (1969).
4. G. Eisenman, ed., Glass Electrodes for Hydrogen and Other Cations. Marcel Dekker, Inc., New York, 1967.
5. H. S. Harned and B. B. Owen, Physical Chemistry of Electro-lytic Solutions, 3rd ed. Reinhold Publ. Corp., New York, 1958.
6. J. N. Butler, Anal. Chem. 39, 1799 (1967); J. Electrochem. Soc. 115, 445 (1968).
7. E. Pungor, K. Tóth, and A. Hrabéczy-Páll, Pure Appl. Chem. 51, 1915 (1979).
8. J. N. Butler, in Ion-Selective Electrodes (R. A. Durst, ed.) NBS Spec. Publ. 314, Washington, 1969, p. 143.
9. H. S. Harned and R. W. Ehlers, J. Am. Chem. Soc. 55, 2179 (1933).
10. M. M. Shultz and A. E. Parfenov, Vestn. Leningrad Univ.,Ser. Fiz. i Khim. 13, 118 (1958).
11. E. W. Moore and J. W. Ross, Science 154, 1553 (1966).
12. A. H. Truesdell, Science 161, 884 (1968).
13. P. Schindler and E. Wälti, Helv. Chim. Acta 51, 539 (1968).
14. R. A. Robinson and R. H. Stokes, Electrolyte Solutions, 2nd rev. ed. Butterworths, London, 1970, appendix 8.10.
15. R. G. Bates and M. Alfenaar, in Ion-Selective Electrodes (R. A. Durst, ed.) NBS Spec. Publ. 314, Washington, 1969, p. 191.
16. J. B. Macaskill, M. S. Mohan, and R. G. Bates, Anal. Chem. 49, 209 (1977).
17. G. Scatchard, J. Am. Chem. Soc. 83, 2636 (1961).
18. K. S. Pitzer, J. Phys. Chem. 77, 268 (1973).
19. P. J. Reilly, R. H. Wood, and R. A. Robinson, J. Phys. Chem. 75, 1305 (1971).
20. R. A. Robinson and V. E. Bower, J. Res. Natl. Bur. Stand. 70A, 313 (1966).

21. J. N. Butler and R. Huston, J. Phys. Chem. 71, 4479 (1967).
22. R. D. Lanier, J. Phys. Chem. 69, 3992 (1965).
23. G. A. Rechnitz and S. B. Zamochnick, Talanta 11, 979(1964).
24. R. D. Lanier, J. Phys. Chem. 69, 2697 (1965).
25. G. Eisenman, in Advances in Analytical Chemistry and Instrumentation (C. N. Reilley, ed.), Vol. 4. Interscience, New York, 1965, p. 213.
26. J. J. Lingane, Anal. Chem. 40, 935 (1968).
27. L. M. Mukherjee, Electrochim. Acta 22, 1255 (1977).
28. L. M. Mukherjee, Electrochim. Acta 17, 965 (1972).
29. T. Nakamura, Bull. Chem. Soc. Japan 48, 2967 (1975).
30. R. Yang, J. Demirgian, J. F. Solsky, E. J. Kikta, Jr., and J. A. Marinsky, J. Phys. Chem. 83, 2752 (1979).
31. G. Akerlof, J. Am. Chem. Soc. 52, 5353 (1930).
32. D. Feakins and co-workers; see for example D. Feakins and P. J. Voice, J. Chem. Soc., Faraday Trans I 68, 1390 (1972).
33. K. K. Kundu, A. K. Rakshit, and M. N. Das, Electrochim. Acta 17, 1921 (1972).
34. C. Treiner, J. Chim. Phys. 70, 1183 (1973).
35. Y. Pointud, J. Juillard, J. P. Morel, and L. Avedikian, Electrochim. Acta 19, 229 (1974).
36. R. Smits, D. L. Massart, J. Juillard, and J. P. Morel, Electrochim. Acta 21, 425 (1976).
37. R. Smits, D. L. Massart, J. Juillard, and J. P. Morel, Electrochim. Acta 21, 437 (1976).
38. H. P. Bennetto and D. Feakins, in Hydrogen-Bonded Solvent Systems (A. K. Covington and P. Jones, eds.) Taylor and Francis Ltd., London, 1968, p. 235.
39. A. K. Das and K. K. Kundu, J. Chem. Soc., Faraday Trans I 70, 1452 (1974).
40. J. F. Coetzee and W. K. Istone, Anal. Chem. 52, 53 (1980).
41. O. Popovych, in Treatise on Analytical Chemistry (I. M. Kolthoff and P. J. Elving, eds.), 2nd ed., Part I, Vol. 1, John Wiley and Sons, New York, 1978, p. 711.
42. K. Srinivasan and G. A. Rechnitz, Anal. Chem. 40, 509 (1968).
43. G. A. Rechnitz and Z. F. Lin, Anal. Chem. 40, 696 (1968).
44. G. A. Rechnitz and T. M. Hseu, Anal. Chem. 41, 111 (1969).
45. K. Tanaka and R. G. Bates, unpublished measurements.
46. R. G. Bates and E. B. Metzer, J. Phys. Chem. 65, 667 (1961).
47. S. P. Datta, A. K. Grzybowski, and B. A. Weston, J. Chem. Soc., 792 (1963).
48. M. Woodhead, M. Paabo, R. A. Robinson, and R. G. Bates, J. Res. Natl. Bur. Stand. 69A, 263 (1969).
49. M. Yoshio and R. G. Bates, unpublished measurements.
50. R. G. Bates and E. B. Hetzer, J. Phys. Chem. 66, 308 (1962).
51. R. G. Bates and G. D. Pinching, J. Res. Natl. Bur. Stand. 42, 419 (1949).
52. M. C. Cox, D. H. Everett, P. A. Landsman, and R. J. Munn, J. Chem. Soc. (B), 1373 (1968).
53. E. B. Hetzer, R. A. Robinson, and R. G. Bates, J. Phys. Chem. 66, 2696 (1962).
54. R. G. Bates and K. Tanaka, J. Solution Chem., in press.
55. P. K. Glasoe and F. A. Long, J. Phys. Chem. 64, 188 (1960).
56. V. Gold and B. M. Lowe, Proc. Chem. Soc., 140 (1963).
57. A. K. Covington, M. Paabo, R. A. Robinson, and R. G. Bates, Anal. Chem. 40, 700 (1968).

58. R. Gary, R. G. Bates, and R. A. Robinson, J. Phys. Chem.
68, 3806 (1964).
59. P. M. Shanbhag and R. G. Bates, unpublished measurements.

Table 1. Comparison of activity coefficients for NaCl in water determined with the Na-ISE (γ_a) with those derived from vapor pressures (γ_b) at 25°C

m	γ_a	γ_b	m	γ_a	γ_b	m	γ_a	γ_b
0.1361	0.754	0.758	0.5273	0.678	0.679	1.0135	0.657	0.657
0.1972	0.733	0.736	0.5765	0.674	0.674	1.0390	0.654	0.656
0.2402	0.716	0.724	0.5997	0.675	0.673	1.0978	0.653	0.656
0.2610	0.714	0.719	0.6439	0.671	0.670	1.1605	0.652	0.655
0.2999	0.708	0.710	0.7089	0.666	0.666	1.2687	0.655	0.654
0.3213	0.703	0.706	0.7656	0.663	0.663	1.3363	0.656	0.654
0.3797	0.695	0.696	0.7906	0.664	0.663	1.4465	0.655	0.655
0.3851	0.695	0.695	0.8643	0.658	0.660	1.6547	0.656	0.657
0.4515	0.685	0.686	0.8993	0.659	0.659	2.0280	0.667	0.668
0.4528	0.685	0.686	0.9161	0.658	0.659	2.0965	0.668	0.671
0.5129	0.679	0.680	0.9790	0.657	0.657	2.1790	0.672	0.674

Table 2. Comparison of the Gibbs energies of transfer for the alkali halides from measurements with cation-selective electrodes with data from amalgam electrodes (in parentheses); data in cal mol^{-1}, molality scale

Solvent	LiCl	NaCl	KCl	RbCl	CsCl	Ref.
20% MeOH	550	820	850	870	810	35
	(580)	(840)	(835)	(820)	(765)	32
					(875)	32
		(825)	(840)			31
30% Ethylene Glycol	570	685	675			35
	(640)	(600)	(650)			33
20% Dioxane	650	810		780		36
	(595)	(760)		(790)		38
60% DMSO	1185	2320	2630	2650	2305	37
	(1240)	(2375)	(2570)	(2615)	(2535)	39

Table 3. pK$_a$ of protonated bases in water from measurements with the Na-ISE (data of Tanaka and Bates) compared with values based on other thermodynamic methods (in parentheses)

Base	15°C	20°C	25°C	30°C	35°C	Ref.
Tris[a]	(8.362)	(8.214)	(8.075)	(7.934)	(7.803)	46
2-Aminopyridine	6.94 (6.949)	6.82 (6.841)	6.73 (6.739)	6.64 (6.639)	6.55 (6.543)	49
Bis	9.10 (9.105)	8.95 (8.951)	8.79 (8.801)	8.66 (8.659)	8.52 (8.519)	50
Ammonia	- -	- -	9.25 (9.245)	- -	- -	51
n-Butylamine	10.93 -	10.79 (10.812)	10.65 (10.640)	10.48 (10.471)	10.30 -	52
sec-Butylamine	10.90	10.73	10.58	10.41	10.24	
t-Butylamine	11.05 (11.048)	10.86 (10.862)	10.68 (10.685)	10.48 (10.511)	10.32 (10.341)	53

, [a] Reference base

Table 4. Comparison of pK$_a$ values for protonated Bis in 50 wt. % H$_2$O/MeOH from Cells G and A. Data of Tanaka and Bates

	15°C	20°C	25°C	30°C	35°C
Tris[a]	8.113	7.963	7.818	7.681	7.550
Bis, cell G	8.74	8.58	8.42	8.27	8.16
Bis, cell A [b]	8.733	8.573	8.423	8.273	8.132

[a] Reference base (48). [b] Ref. (54).

Table 5. pK$_a$ for protonated Tris in D$_2$O (45)

	15°C	20°C	25°C	30°C	35°C	40°C
Cell B, phosphate standard	8.979	8.825	8.678	8.526	8.391	8.267
Cell G, calculated E°	-	-	8.678	-	-	-

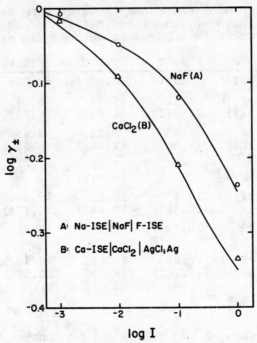

Figure 1.   Activity coefficients of NaF and CaCl$_2$ at 25°C
            from cells with ion-selective electrodes, as a
            function of ionic strength.
               Circles and triangles, experimental points
               Solid lines, literature data

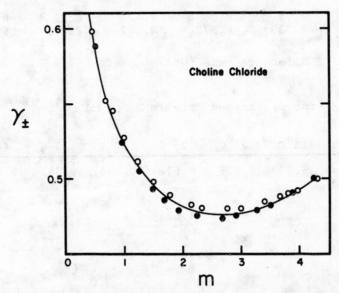

Figure 2.   Activity coefficients of choline chloride at 25°C
            from measurements with an ion-selective electrode
            in cell E (dots) compared with data from isopiestic
            vapor pressure measurements (circles)

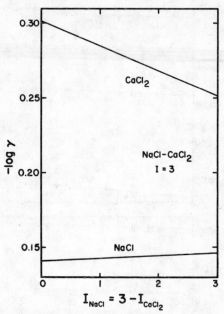

Figure 3. Variation of the activity coefficients in mixtures of NaCl and CaCl₂ (I=3.0) at 25°C with composition of the salt mixture

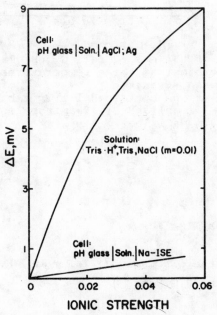

Figure 4. Variation of the emf of cell B (upper curve) and cell G (lower curve) with changes in ionic strength
Both cells contain Tris·H⁺(I-0.01),Tris(I-0.01), and NaCl (0.01m)

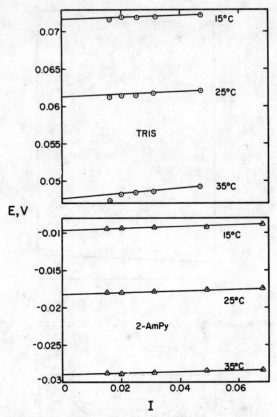

Figure 5. Plot of the emf of cell G (pH glass and Na-ISE
electrodes) at three temperatures as a function
of ionic strength
Solutions contained 1:1 Tris buffer in 0.01m
NaCl (top) and 1:1 2-aminopyridine buffer in
0.01m NaCl (bottom)

QUESTIONS AND COMMENTS

Participants of the discussion: E.Pungor, K.Burger, É.Deák,
W.Simon, R.P.Buck, R.G.Bates

Question:
Is it justified to give pK values in four digits ? Were the
results analyzed with respect to accuracy ?

Answer:
The measurements were made with a pH and a sodium glass
electrode. I do not think that by this way a greater accuracy
than ± 0.01 to 0.05 pH or pNa can be expected. That would be
a reasonable estimation.

Question:
Some activity coefficient data were presented in your lecture,
which were measured in mixed solvents with a fairly low dielectric constant. Was ion-pair formation taken into account
in calculating the activity coefficients ?

Answer:
The data shown for mixed solvents were pK data measured in
50% methanol. I do not think we have to worry about ion-pair
formation.

Question:
What was the highest concentration you had ?

Answer:
During pK measurements the highest concentration was around
an ionic strength of 0.06.

Question:
How was the Nernstian behaviour of the ion-selective electrodes
used checked ? Were only the methods presented used, or did
you use also some additional ones ?

Answer:
Actually, our results confirmed the fact that the Nernst slope
was valid because our results have been compared with thermodynamically valid and accepted data, and the sodium glass electrode and the pH electrode reproduced these values from 20
to $85^{\circ}C$ so I think we have at least indirectly confirmed the
validity of the Nernst slope.

Question:
Is it correct that TRIS has a pK value of about 8 in water at
25°C, and BIS about 8.6 which means an increase of 0.6 pK
units on replacing the CH_2OH group by a CH_3 group ?

Answer:
The pK of TRIS is 8.07 in water at 25°C, and that of BIS is
about 8.8 roughly. There is some indication that the isopiestic effect is a linear function of pK itself, but it has
not been proved yet.

Question:
You get these activity coefficients in mixed solvents from
isopiestic data. Why are so few of these known and why is it
so difficult to test them ?

Answer:
Most of the work reported here has been done with e.m.f.
measurements where the activity coefficients of hydrochloric
acid were determined. The determination of activity coefficients
from isopiestic measurements is perfectly possible thermodynamically. It depends upon knowing the activity coefficients
of each of the electrolytes alone, which can of course be
obtained by isopiestic measurements and then making measurements on mixtures of a constant composition with respect to
sodium chloride and varying the calcium chloride concentration.
All go back to the Gibbs-Duhem equation which can be integrated for these mixtures.

Question:
It was somewhat surprising to hear that high concentrations
are used to measure activity coefficients by vapour pressure
osmometry. It was mentioned that this is due to the lack of
sensitivity at lower concentrations, but this has changed
dramatically in the past few years. There are instruments
available now by which one can measure in 10^{-4} M solutions
with a reproducibility of about 1%. So, would it not be

24

interesting to use this method for activity coefficient determinations ?

Answer:
Yes, we were very much interested in these instruments, the
"Baratrons". In our experience, however, the reproducibility
you mentioned is very difficult to achieve at vapour pressure
measurements at very low concentrations. You have to degas
the system for a great length of time. I was referring to the
isopiestic vapour pressure measurement which is of course
relative, and is extremely convenient. But if you try to
equilibrate a solution of sodium chloride with a solution of
calcium chloride in a vacuum desiccator, it takes days and days
and perhaps weeks if the concentration is below an ionic
strength of 0.1. So I think that these instruments do have a
great promise but so far they have not been used to obtain
accurate data as far as I am aware.

Comment:
Weisenberg at the McGillan University is an expert on the
disposition of water of iron/II/ and iron/III/ in polymers.
There is now a tremendous evidence of a new membrane called
"Nephiron", which is an ethylenetetrafluoro polymer with
sulphonic acid groups and it is very clear that there are
hydrophobic regions, and from the infrared spectra you get the
average clustering in water and from the Mössbauer spectra
you get the number of free ions, the number of dimers and
number of clusters. The trouble with this is about the
conductivity measurements because we do not know the concentrations of the charge carrying species, so we get the so
called gross mobility which may be much larger because there
are many actual carriers.

It is a very interesting piece of work, he measures the
glass transition points of polymers, and he has another way of
studying the physical chemistry of polymers.

THE IMPEDANCE METHOD APPLIED TO THE INVESTIGATION
OF ION-SELECTIVE ELECTRODES

RICHARD P. BUCK

Department of Chemistry, The University of North Carolina, Chapel Hill, N. C. 27514, USA

AIM OF THE METHOD

The direct purpose of the impedance experiment is determination of a descriptive system function for ion selective electrodes (ISEs). The impedance function relates output voltage to perturbing input current. The impedance is a function of the perturbing electrical frequency, temperature, thermodynamic and kinetic parameters, and bathing activities. The indirect purpose of the method is determination and interpretation of the time constants of ISE responses.

The impedance is one of several functions that can be measured and used to characterize electrical passive networks. It is an easily determined function that requires only two terminals and a reference ground potential. The inner and outer reference electrodes of a typical ISE cell provide the two terminals: an input and an output. However, when the membrane has to be changed, more convenient cells can be built. For example, Figure 1 gives a cell used for liquid ion exchanger membrane studies. By data treatment, the measurement gives two other interesting properties: 1) the current through the electrodes vs time after application of a voltage step across the cell, or 2) the voltage between the electrodes vs time after application of a current step.

Most important for the study of ISEs is interpretation of experimental electrical responses in terms of predicted responses from models of interfacial and bulk, charged-species transport. If a model is correct, an interpretation of time constants can be made using independently measured properties of the membrane electrode and cell system. The impedance method also provides one of several alternative methods for identifying the membrane properties responsible for the potentiometric responses. Models that predict correctly the potential differences across ISE membrane electrodes must also give logical interpretations of time responses after voltage and current pulses and steps.

Correlations of measured parameters like resistivity and capacity with electrode thickness and area provide partial tests. Identification of surface resistance that correlates with slope deviations from the equilibrium Nernst values is another test. The chief flaw in the method is its gross nature. The direct results of impedance measurements are electrical equivalents; R, C, and L. Interpretations are possible only in macroscopic terms. It is not possible to deduce microscopic aspects of transport or

27

double layer structure from impedance measurements alone.

ASSUMPTIONS

Application of the impedance method requires an assumption that real ISEs
in cells correspond electrically to two-port passive networks composed of
lumped parameters; resistors, capacitors and inductors. There is no re-
quirement that the equivalent network be finite, and infinite chains of RC
elements sometimes occur in networks equivalent to diffusion potentials,
e.g. the analog for the Warburg process. It is likely that all distributed
electrochemical systems in the fields of sensor, battery, dielectric and
corrosion technologies have equivalent network analogs. Since the materials
are composed of ionic, electronic and mixed conductors contacted by metals,
it is believed that each system will have one or more network equivalents
which behave identically under a small perturbation. Another way of speaking about the equivalent systems is to say that both are described by i-
dentical linear differential equations. Solutions give typically $\phi_{out}(t)$
or $i_{out}(t)$ for input excitations $i_{in}(t)$ or $\phi_{in}(t)$, where t is time, ϕ is
potential difference or measured voltage, and i is current.

THE IMPEDANCE 'THOUGHT' EXPERIMENT

A two-port passive network may be characterized by applying an input current
or voltage signal and noting the output voltage or current. When i_{in} =
$i(\omega t)$ is a sin or cos function, then ϕ_{out} = $\phi(\omega t)$. In Figure 2, for a
driven input sinusoidal current of frequency

$$i(\omega t) = i° \sin (\omega t) \tag{1}$$

the output voltage is most generally

$$\phi(\omega t) = Hi° \sin (\omega t + \theta) \tag{2}$$

If the driving current is chosen small enough that the coefficient Hi° is
a voltage amplitude less than RT/F (23mV at 25C), the ISE will respond according to a linear differential equation. The current amplitude multiplier
H is found to be

$$H = (Z_R^2 + Z_I^2)^{1/2} \tag{3}$$

and

$$\theta = \tan^{-1} (\frac{-Z_I}{Z_R}) \tag{4}$$

H is absolute value of the impedance, Z_R and Z_I are real and imaginary impedance components and θ is the phase angle shift. There are no higher
overtones, e.g. no components of $2\omega t$, $3\omega t$, etc.

Since any $\sin (\omega t + \theta)$ can be decomposed into an 'in phase' real component,
$\sin \omega t$, and an 'out-of-phase', imaginary component, $\sin (\omega t + \pi/2)$ or $\cos$
ωt, the response voltage can, just as well, be written

$$\phi(\omega t) = Z_R i° \sin \omega t + Z_I i° \cos \omega t = Hi° \cos \theta \sin \omega t + Hi° \sin \theta \cos \omega t \tag{5}$$

The coefficients are the real and imaginary impedances. Z_R, Z_I and consequently H and θ are, themselves, functions of frequency ω.

Another way of viewing the output is to consider an Argand diagram shown in
Figure 3. The real vector of amplitude Z_R moves back and forth (lengthens
and shortens and reverses) along the real axis, while the imaginary vector

of amplitude Z_I moves similarly on the imaginary axis. The actual vector lies at an angle θ with real axis and rotates counterclockwise with ω.

A more general approach to the impedance function is the application of the fourier transformation. Given $i(t)$ input and $\phi(t)$ output,

$$\int_0^\infty e^{-j\omega t}\phi(t)dt = Z(j\omega) \int_0^\infty e^{-j\omega t}i(t)dt \qquad (6)$$

or

$$\bar{\phi}(j\omega) = Z(j\omega)\bar{i}(j\omega) \qquad (7)$$

which bears a formal resemblance to Ohm's Law. The impedance function $Z(j\omega)$ is a complex function with the previously identified parts

$$Z(j\omega) = Z_R + jZ_I \qquad (8)$$

The admittance function $Y(j) = Y_R + jY_I$ relates an output current to an input voltage. The experiments are related because

$$Y_R = Z_R / (Z_R^2 + Z_I^2) \qquad (9a)$$

$$Y_I = -Z_I / (Z_R^2 + Z_I^2) \qquad (9b)$$

THE EXPERIMENT

The classical experimental set-up was a resistance-capacitance (impedance) bridge actuated by an oscillator with controlled, variable frequency. The cell was inserted in one arm of the bridge and a series or parallel combination of standard R and C was substituted until a null was achieved on an oscilloscope. This procedure was repeated at each frequency and $R(\omega)$ and $C(\omega)$ values were tabulated. The impedance function, and real and imaginary parts were then computed from

$$Z(j\omega) = R(\omega) + 1/j\omega C(\omega) \text{ or } Z(j\omega) = R(\omega)/ 1 + j\omega R(\omega)C(\omega) \qquad (10a,b)$$

for series and parallel cases. If the cell were represented by a simple RC circuit, the measured values of R and C were independent of frequency.

Modern impedance experiments use correlation techniques or electromechanical phase angle and absolute impedance measurements. The first type, like a phase-sensitive detector or lock-in amplifier, measures amplitudes of the in-phase and out-of-phase components. The second category is already obsolete, but gives two equivalent measurements. From the first:

$$Z_R = \frac{\phi_{Real}}{i_{in}} \ ; \ Z_I = \frac{\phi_{imag}}{i_{in}} \qquad (11a,b)$$

and from the second

$$Z_R = Z_T \cos \theta; \ Z_I = Z_T \sin \theta \qquad (12a,b)$$

A BASIC IMPEDANCE FUNCTION AND THE IMPEDANCE PLOT

Consider an impedance measurement on a simple parallel RC network in Figure 4. In this case

$$Z(j\omega) = \frac{R}{1 + j\omega RC} \qquad (13)$$

Impedances add like resistances using functions

$\quad$ R = Resistance
$\quad$ $1/j\omega C$ = Reactance for C
$\quad$ $j\omega L$ = Reactance for L

For the parallel network, the components are added as reciprocals. The real and imaginary parts are:

$$Z_R = \frac{R}{1 + \omega^2 R^2 C^2} = \frac{R}{1 + (\omega\tau)^2} \qquad (14a)$$

$$Z_I = \frac{-\omega R^2 C}{1 + \omega^2 R^2 C^2} = \frac{-\omega\tau R}{1 + (\omega\tau)^2} \qquad (14b)$$

with τ = RC the time constant. The dimensions of ω are s^{-1} (radians/sec); R (volt sec/coul); C (coul/volt); RC(s). Consequently $\omega\tau$ is dimensionless. A plot of log Z_R vs. log ω is horizontal at $\omega\tau<1$, but declines with slope -2 for $\omega\tau>1$. Log Z_I vs. log ω is a peaked function with slope +1 for $\omega\tau<1$ and slope -1 for $\omega\tau>1$. The peak occurs at $\omega\tau = 1$.

The Nyquist complex plane plot is parametric in ω. $-Z_I$ is plotted against Z_R. This plot is a semicircle as shown in Figure 5a. Intercepts on the real axis occur at $\omega = \infty$ and $\omega = 0$. The maximum occur at $Z_R = R/2$ where $-Z_I = R/2$ and $\omega\tau = 1$. The circle (with center at 0, R/2) can be demonstrated by noting that

$$(Z_R - R/2)^2 + Z_I^2 = R^2/4 \qquad (15)$$

In Figure 5a, the dc resistance R_0 is identified as R_∞ because the left hand semicircle in any impedance plot of a real electrode is the so-called high frequency or infinite frequency response. In this circuit R_∞ and R_0 are the same. The equivalent conductance G_∞ is also used. (1,2)

Two other simple circuits are illustrated in Figure 5b and 5c using both impedance plane and admittance plane plots. These Nyquist plots are sometimes called Cole-Cole plots, but they predate the Coles' work (3), and they are not the same in detail. Cole-Cole plots treat the measurement data by an equivalent leaky capacitor circuit. Consequently their method plots real and imaginary dielectric constant. There is great superficial similarity in the diagrams, however.

Impedance plane plots are not always semicircles (1,2,4). They will show vertical lines, ideally, for systems with capacitive coupling at dc as in Figure 5b. The semicircles can be distorted for three main reasons: 1) Two semicircles are overlapping, because two parallel RC networks in series have similar time constants, differing by less than a factor of 10. 2a) Many parallel RC elements occur in series. When the time constants are nearly randomly distributed about a mean, the semicircle appears 'sunken' e.g. the center lies below the real axis. (4) The general expression is an extension of Eq. 10b.

$$Z(j\omega) = \frac{R}{1 + (j\omega\tau)^\alpha} \qquad (16)$$

2b) Many parallel RC elements occur in series in a way equivalent to a transmission line with equal C and R per unit length. This behavior is recognized as a diffusional Warburg impedance (5). The impedance diagram Figure 6b (right side) shows an example.

SUPERFICIAL STRUCTURE OF MEMBRANE CELLS

Consider an ordinary ISE cell:

$$Cu \mid Ag \mid AgCl \mid K^+Cl^- \mid \begin{matrix}test \\ solution\end{matrix} \mid membrane \mid \begin{matrix}inner \\ filling \\ soln\end{matrix} \mid AgCl \mid Ag \mid Cu \qquad (17)$$

The two copper wires are the input and output terminals. The silver wires and AgCl coating should possess sufficient area that the interfacial resistance is low and capacitance is large in comparison with the same parameters of the membrane. In addition stirred solution resistance should be less than the membrane resistance. In most cases, circuit impedances in series with membrane impedances are small (10^2 to 10^4 Ω in comparison with 10^5 to 10^7 Ω for most ISEs). Membranes used in ISEs are 'thick' in the sense they are electrically neutral in the interior, and their resistances rarely exceed 10^7 Ω in the case of glass electrodes. The cell with stirred solutions should be tested, when feasible, without the membrane, to determine that the residual impedance is 1% or less of the membrane impedance. When membrane impedances are comparable with cell impedances without membrane, impedance plane plots are more difficult to interpret.

The ISE impedance can be primarily determined by the membrane itself: Factors that determine equilibrium and steady-state membrane impedances at room temperature are (6):

1. Chemical homogeneity on a microscopic level.
2. Physical uniformity and freedom from cracks, grain boundaries or other unusual transport pathways.
3. Number and kind of charge carriers.
4. Mobility of charge carriers
5. Rate of transfer of carriers from electrolyte.
6. Rate of adsorption when adsorption-reaction paths are involved.
7. Rate of generation and recombination of charge carriers from complexes, ion pairs and lattice sites.
8. Location and types of space charge, adsorbed charge and surface ionic states.

EQUIVALENT CIRCUITS FOR ISEs

The theories of transport and space charge distributions give locations and kinds of resistance and capacitance. Fortunately for thick ISE membranes, the corresponding time constants do not often overlap. This circumstance means processes involving coupled R and C fortuitously allow RC products to be separated by factors of 10^2 or more. The identities of the processes are summarized in Table I. These time constants in order of increasing magnitudes are:

$\tau_B = R_B C_B$ bulk, geometric (1) (18a)

$\tau_R = R_R C_R$ surface activation (7) overpotential (18b)

$\tau_F = R_F C_F$ resistive surface film or resistive region inside surface (4) (18c)

$\tau_A = R_A C_A$ coupled adsorption, reaction ion transfer (8) (18d)

$\tau_G = R_G C_G$ carrier generation recombination (9) (18e)

$\tau_W = R_W C_W$ Warburg diffusion of charge carriers and complexes (1,2,5,10) (18f)

The expected impedance plot is shown in Figure 6 for well-separated or

31

loosely coupled processes. When the processes overlap, the semicircles move together, and the imaginary impedance does not decrease to zero between semicircles. For each semicircle, the frequency at the top (maximum - Z_I) obeys

$$\omega_{max}\tau = 1 \qquad (19)$$

from which τ can be computed. For distorted semicircles, such as the Warburg impedance, finding τ_{max} is not difficult, but identifying the corresponding R_W and C_W is difficult. R_W is called R_D in Figure 6, because diffusion is controlling at dc. From each semicircle, the product RC is found directly. Then the width on the real axis is the corresponding R. Capacitance C is calculated from τ and R.

In Figure 6 and in the listing of time constants, only those associated with equilibrium or steady state ISEs were considered. If the electrolyte solution resistances are comparable with the membrane resistance R, then the infinite frequency intersection on the real axis will not occur at zero ohms, but at a measurable value greater than zero. An unlikely situation for ISEs, but one which must be faced in dealing with some thin films, is the possibility that the membrane resistance is, in fact, less than the solution resistance. The result can be exterior concentration polarization (depletion) of charge carriers on one side shown in Figure 7. In an unstirred bathing solution, the result is another Warburg impedance at lower frequencies. Rapidly stirred solutions compress this Warburg behavior by thinning the exterior Nernst boundary layer.

EXPECTED EQUIVALENT CIRCUITS FOR ISEs.

ISEs that function rapidly are based on reversible ion exchanging chemical systems. One would expect to see impedance functions and plots (and voltage-time responses) that are simple. Perhaps one or two, at most three, semicircles might be encountered for smooth homogeneous electrodes. For the ϕ_{out} vs time response, one to three exponentials are expected. Experimentally simple plots are encountered (Figure 8). The great variety of processes predicted by theory have not been seen for conventional glass, crystalline and non-crystalline ISEs such as Ag_2S shown in Figure 8. Responses in the megaHertz range have not been measured, and so factors such as dipole relaxation in liquids and defect motion in solids have not been found. Rather, the main features of practical ISE behavior in the time or impedance plane, have been τ_B, τ_R or $_F$ and τ_W (for liquids).

For simple, liquid or solid, homogeneous systems and well-characterized complex systems, such as homogeneous glass electrodes with hydrolyzed film layers and localized resistive layers, time constants can be identified and interpreted. The τ_B semicircle is expected for all membrane systems contacted by electrolytes or by metallic conductors. A τ_R semicircle is a sign of irreversibility, and the corresponding activation overpotential has been found for sulfide electrodes and some nitrate electrodes (see later). More commonly, resistive surface layers exist, and ions from the electrolyte must penetrate this region. The resulting τ_F is distinguishable from τ_R by etching the surface, for example, but τ_F has the same consequences: a surface region resistance. Warburg τ_W is not expected for homogeneous solid electrodes or glasses with one charge carrier. τ_W is expected whenever two charge carriers of differing mobilities are involved in the conduction. This situation can arise in bathing electrolytes at electrode surfaces, in crevices and along grain boundaries in non-homogeneous solids and, especially, in liquid ion exchanger membranes. Thus, liquid electrodes inevitably show a diffusion (concentration polarization) impedance at low frequencies, while single crystals do not, but pressed pellets may show the effect. The complete Warburg response is not always observable because of frequency limits of the instrument or internal stirring in the membrane. Long time drift responses of ϕ vs t are seen with alkali ion-sensing glass electrodes, and occasionally with crystalline electrodes. These results are difficult to confirm by observing a Warburg impedance. Processes requiring 10 s or more to complete are not readily observed by ac measurements.

Long time constants have been observed for solid membranes with cracks or grain boundary transport. When parallel processes can occur, (e.g. bulk and grain boundary transport) one expects two semicircles to show up depending on extent of overlap. Some unexplained complexities involving semicircle overlap have been observed for non-membrane systems, such as electrodes of the second kind when the surface materials are porous. (11,12)

SELECTED EXAMPLES

Glass

Sandifer and Buck (4) extended to lower frequencies, measurements of two Beckman pH glass electrodes previously studied by Buck and Krull (13). One of the glasses, so-called "general purpose" pH glass was measured, in electrode configuration and found to show a two-semicircle impedance plane plot in Figure 9. The frequency range was 2kHz to 6×10^{-3} Hz for a temperature range 5 - 65C. Lowering of the semicircles because of a distribution of time constants has been interpreted to mean that the surface is rough, possibly with microcracks, and that many diffusion-migration pathways are possible. When the resistances and corresponding capacitances are not geometry-sensitive in exactly the reciprocal way, then a nearly-Gaussian distribution of time constants centered on the normal geometric τ_B, results. The individual $R(\omega)$ and $C(\omega)$ plots (Figures 10a, b) are shown to illustrate how much simpler the impedance plane plots are for interpretation. Transport enthalpy for the bulk glass was 0.668 eV., but the surface region activation enthalpy was 1.11 eV.

To demonstrate the low frequency semicircle dependence on outer layer properties, the glass was successively etched and the impedances remeasured. The bulk semicircle did not change, but the reaction semicircle decreased as shown in Figure 11.

Silver chloride

Impedances were measured on purified silver chloride crystals of thickness 0.064 cm to 0.400 cm at temperatures 20 - 35 C (14). Although the single crystals were purified, residual divalent heavy metal impurities conferred vacancy conductivity. Results showed one semicircle, independent of solution contact (Ag^+, Cl^-, or NO_3^-) in Figure 12a. Ion exchange processes are reversible for the frequency range 10 kHz to 0.01 Hz. When solid contacts were used, it was difficult to reproduce exactly the contact area of the solution. Thus, dc intercept varies as shown in Figure 12b. Nevertheless, only the bulk semicircle appears. Slow ion exchange kinetics can be forced to occur by using organo-silver paint electrodes as illustrated in Figure 13. Normally Hanovia silver contacts are formed by baking out the organic residues. However, AgCl crystals will not tolerate the necessary high temperature.

The electrical relaxation-time τ_B was 0.195 ± 0.021 ms for these crystals and was independent of crystal thickness and bathing solution composition, but was dependent on temperature. The resistive component of the impedance R_∞, was linear with crystal thickness, δ and the geometric capacitance C_g linear with δ^{-1} as shown in Figures 14a and b. R_∞ is equal to dc resistance, R_0. This indicates in addition to rapid ion exchange at the surface, no Warburg diffusion. The dimensionless dielectric constant, calculated from C_g and the crystal dimensions, is 15.4. An average activation energy of 0.298 eV, indicative of vacancy transport is found. The parameter α in the equivalent non-ideal Cole-Cole representation is 0.96, indicating a narrow distribution of time constants as expected for physically homogeneous single crystals.

Silver bromide

Impedance measurements were also made on purified, vacancy-doped and inter-stitial-oped silver bromide crystals of thicknesses from 0.061 cm to 0.260 cm using constant ionic strength solution contacts at temperatures from 20 C - 35 C (15). Results characterize the macroscopic transport properties from 316 kHz to 0.01 Hz and at dc. One semicircle is observed. In all cases the high frequency limit resistive component of the impedance, R_∞, is equal to the dc resistance R_0. This result indicates rapid ion exchange at the surfaces and neither Warburg diffusion nor surface reaction, kinetic limitation. As expected, R_∞ is a function of charge carrier concentration, which depends on extrinsic dopant concentration over the temperature range measured. For pure crystals the resistance, R_∞ is linearly dependent on crystal thickness, δ, while thickness correlation could not be tested for the doped crystals as dopant level was not uniform. The geometric capaci-tance for all crystals, C_g, is linear with δ^{-1}. The dimensionless dielectric constant, κ calculated from C_g and crystal dimensions is 13.9 ± 0.7. The electric relaxation time τ_B at 25 C is found to be 50 ± 3 µs for pure crys-tals, 1 to 4 µs for Cd^{2+} doped crystals and 12 to 66 µs for the S^{2-} doped crystals. Temperature dependance of R_∞ allows determination of crystal transport activation energies. For pure and Cd^{2+} doped crystals a value of 0.33 ± 0.02 eV is found. For the S^{2-} doped crystals values from 0.48 to 0.28 eV are found. The parameter α, in the non-ideal Cole-Cole represen-tation of impedance plane arcs, is 0.96 ± 0.1 for purified crystals, 0.93 ± 0.03 for Cd^{2+} doped and 0.92 ± 0.02 for S^{2-} doped crystals.

Silver sulfide

Silver sulfide single crystals are difficult to grow. Instead, pressed pel-lets were studied using an ohmic silver interior contact. The membranes could be rotated to achieve a stable Nernst diffusion layer at the exterior electrolyte interface. Silver sulfide is a mixed conductor and the ionic resistance is small compared with the halides reported above. This means that higher currents are used in the experiment to achieve a measurable cell voltage. The possibility of exterior electrolyte concentration polarization is recognized and avoided by rotating disk configuration, rapid electrode rotation. (16)

Membranes were prepared from materials precipitated in excess silver or sulfide ions, and from stoichiometric mixtures. In addition to $Ag/Ag_2S/Ag^+$ (aqueous) cells, regular ohmic cell, $Ag/Ag_2S/Ag$, measurements were made. Impedances were determined from 31.6 kHz to 0.0178 Hz. High frequency bulk resistivities, R_∞ varied by four with precipitation and pressing conditions. Resisivities were the same for solution and ohmic configurations for each preparation. For ohmic contact, R_∞ and R_0 were identical. Solution contact cells in 10^{-1} M and 10^{-2} M bathing silver ion solution gave identical frequency-dependent impedances which were independent of rotation rate. Examples of impedance plane plots are shown in Figure 15. These examples are complicated because they are composites of τ_B, τ_W, and τ_R. It was noted that the concentration-invariant shapes could be resolved into τ_B at highest frequencies and τ_W at lowest, for the high concentration 10^{-1} - 10^{-2} M bathing solutions. At low concentrations, 10^{-4}, 10^{-5} M, a further merged semicircle was found, and was interpreted as τ_R. The latter was determined by subtracting out impedances measured in the higher concentration solutions. The method used least square fitting, but cannot be considered as an exact analysis. Estimate of a rate constant 2.1×10^{-4} cm s^{-1} was similar to the value estimated by Cammann and Rechnitz (17) from entirely different experiments involving controlled currents or voltages. As pointed out earlier, the same type of information, if real, must be extractable from all related measurements. The analysis using the circuit of Figure 16 also suggested the possibility of an impedance arising from dissolution of the crystal or soluble coprecipitated components into the bathing solution. This impedance corresponding to the crystallization potential was predicted by other workers, but not easily observed. It is a non-equilibrium effect that can be expected when test solutions are not saturated with membrane components. The parameters computed to give a good fit are collected in Table II.

Liquid Ion Exchangers

Using Aliquat nitrate (tricaprylmethylammonium nitrate) in nitrobenzene solvent, impedances were measured from dc to 5 kHz. Platinum, blocking electrodes at controlled spacing were used to obtain system data on the τ_B semicircle (18). These are shown in Figure 17. Derived quantities are shown in Tables III-V.

The impedances of free-standing liquids held between the hydrophobic polymer sheets, provides different plots as shown in Figure 18a. The polymers are contacted by KNO_3 bathing solutions. The films behave as surface resistances and provide a τ_R semicircle (19). Water permeable membranes as supports give the beginning of a Warburg impedance. An example of free-standing liquid between dialysis films is shown in Figure 18b. Again, KNO_3 bathing electrolytes contact the membranes. In the conventional ISE configuration for liquid ion exchangers, a single microporous membrane is saturated with liquid. If the membrane has defined pores, the liquid is held mainly in the pores. Then the resistance is proportional to the pore cross-sectional areas, since transport via pores is a low resistance pathway. However, nearly homogeneous membranes such a PVC and PVC-acrylonitrile dissolve the nitrobenzene solvent. Then transport occurs throughout the membrane bulk and the enthalpy of transport is greater than that for Aliquate nitrate in nitrobenzene solvent.

A more interesting factor occurs in the PVC-acrylonitrile support-membrane case in Figure 19. Not only are the τ_B, and τ_W regions present, (the latter is only partially formed), but a τ_R sunken semicircle occurs as well. These results are shown over a period of time in Figure 20. The middle semicircle is surface kinetic, because it is independent of support membrane thickness. As the membrane is soaked in electrolyte for many hours, water penetrates into the membrane along strands of the support material. Water does not affect transport parameters of the Aliquat nitrate itself. This fact was noted by comparing dry vs. wet data in Table III. Also, the τ_B remains the same in Figure 20, even though R_∞ is decreasing. As R_∞ decreases and C_g increases, τ_B remains constant only if the transport mobilities remain constant.

INFORMATION OBTAINED FROM IMPEDANCE PLANE PLOTS

Because of limited frequency range of equipment, about 10^{-2} Hz to about 1 MHz, those processes that can contribute to impedances must occur in about $10^{-6}/2\pi$ to $10^2/2\pi$ seconds. Some high frequency processes such as dipole relaxation, and some long time drifts cannot be conveniently studied. But, as illustrated in the examples, many processes can be observed.

From τ_B and R_∞, R_0, a decision on charge carriers can be made. If R_∞ and R_0 are the same, probably only one charge carrier is involved, and thus carriers also cross the interface. If $R_0 > R_\infty$, then fewer charge carriers cross the interface in comparison with those that move under an applied field within the membrane bulk. From τ_B, R_∞, C_g and dielectric constant $\varepsilon = \kappa\varepsilon_0$ can be found. By analysis of R_∞ vs concentration of carriers, salt mobilities, equivalent conductances and ion pairing constants can be found. By studying impedances over a temperature range the activation parameters for transport can be found, and the thermodynamic properties of complexes and ion pairs can be determined.

From τ_R, some conclusions about surface rates, appearing as interfacial resistances can be made. If the apparent rate changes with surface etching or polishing, it is likely that the surface rate occurs because of a high resistance layer. If the rate is the same regardless of surface renewal, then a slow, potential-dependent ion-transfer is indicated.

A Warburg impedance is indicated by a 45° line at the lower frequency side of the diagram as in Figure 6. If the entire distorted semicircle can be observed, then the relative mobilities of unblocked carriers can be determined for single electrolyte systems,

$$R_0 = R_\infty(1 + \frac{\overline{u}_- z_-}{\overline{u}_+ z_+}) \tag{20a}$$

for permeable cations and

$$R_0 = R_\infty(1 + \frac{\overline{u}_+ z_+}{\overline{u}_- z_-}) \tag{20b}$$

for permeable anions.
But as pointed out, in heterogeneous systems such as pressed pellet electrodes, a Warburg impedance from internal concentration polarization of charge carriers can be seen.

CONNECTIONS BETWEEN IMPEDANCE PLOTS AND VOLTAGE-TIME EXPERIMENTS FROM ACTIVITY STEPS

Impedance plots relate total, real or imaginary impedances and frequency. The Fourier transform gives cell voltage vs. time after application of a small perturbing current. The various time constants described are connected to the physical processes for transport of charged species across interfaces and through the bulk of membranes used as ISEs. The act of measuring the cell voltage after an activity step of potential altering magnitude, involves also the same transport processes. However, the processes may not affect the outcome (ϕ vs t) in the same way.

An analysis of this problem was given by Stover, Brumleve and Buck (20). The chief result was an observation that fewer time constants are observed in an activity step experiment compared with the current step experiment. The reason is that forced current affects space charge distributions, electric fields and ionic motion at both interfaces and throughout the membrane. But the activity step affects fields and space charge at one interface. The widest variety of effects come from liquid ion exchanger membranes. In Figure 21 ln-ln plots of reduced responses vs. time are illustrated for a single counter ion bathing solutions with equilibrated membrane. A 1% activity steps on one side (curve a) and a current step (curve b) to create identical equilibrium potential changes show the same geometric time constant (τ_∞), but only curve b shows the Warburg τ_W. The reason is that the current step requires readjustment of site concentration, while the activity step does not.

In Figure 22, an equilibrated single counter ion system is replaced on one side by another counter ion (e.g. KCl replacing NaCl at a cation-permeable membrane). Curve a is the result when the two counter ions have different mobilities (thus requiring site motion and τ_W); curve b shows the result when counter ions have the same mobilities (no site readjustment required and no τ_W). Of course, τ_∞ occurs in both.

In Figure 23, an activity step (curve a) and an equivalent current step (curve b) are compared for the case of a single counter ion with a slow surface rate. Note the absence of a charging time constant τ_∞ and the dominant effect of the slow surface rate in both cases. Thus slow interfacial kinetics limit responses after the activity step. The bulk semicircle charging processes do not occur on the expected rapid time scale.

REFERENCES

1. J. R. Macdonald, J. Chem. Phys. 61, 3977 (1974).
2. J. R. Macdonald, Electroanal. Chem. and Interfac. Electrochem. 53, 1 (1974).
3. K. S. Cole and R. H. Cole, J. Chem. Phys. 9, 341 (1941).
4. J. R. Sandifer and R. P. Buck, J. Electroanal. Chem and Interfac. Electrochem. 56, 385 (1974).
5. R. P. Buck, Crit. Rev. Anal. Chem., CRC Press, 5, 323 (1975).
6. J. R. Macdonald, "Interpretation of AC Impedance Measurements in Solids", Superionic Conductors, G. D. Mahan and W. L. Roth (eds.) Plenum Pub. Corp. (1976), pg. 81.
7. D. R. Franceschetti and J. R. Macdonald, J. Electroanal. Chem. and Interfac. Electrochem. 82, 271 (1977).
8. J. R. Macdonald, ibid. 70, 17 (1976).
9. J. R. Macdonald and D. R. Franceschetti, J. Chem. Phys. 68, 1614 (1978).
10. R. P. Buck, J. Electroanal. Chem. and Interfac. Electrochem. 18, 381 (1968).
11. R. K. Rhodes and R. P. Buck, Anal. Chim. Acta, 113, 55 (1980).
12. R. K. Rhodes and R. P. Buck, Anal. Chim. Acta, 113, 67 (1980).
13. R. P. Buck and I. Krull, ibid 18, 387 (1968).
14. R. P. Buck, D. E. Mathis and R. K. Rhodes, ibid, 80, 245 (1977).
15. R. K. Rhodes and R. P. Buck, ibid 86, 349 (1978).
16. R. K. Rhodes and R. P. Buck, Anal. Chim. Acta, 110, 185 (1979).
17. K. Cammann and G. A. Rechnitz, Anal. Chem. 48, 856 (1976).

18. D. E. Mathis and R. P. Buck, J. Memb. Sci. <u>4</u>, 379 (1979).
19. D. E. Mathis, F. S. Stover and R. P. Buck, ibid, <u>4</u>, 395 (1979).
20. F. S. Stover, T. R. Brumleve and R. P. Buck, Anal Chim. Acta, <u>109</u>, 259 (1979).

TABLE I.   PROCESS IDENTIFICATION

τ	R	C
B: Bulk, geometric	R_B = R, resistance for uniform carrier concentration. All charged membrane species carry current, regardless of blocking at interfaces.	$C_B=C_g$, geometric capacitance of both electrolyte space charge regions at outer surfaces of membrane (picofarads/cm^2)
R: Reaction, surface activation over-potential	R_R surface resistance equivalent to Butler-Volmer-Erdey Gruz activation overpotential required to move an ion across the phase boundary.	$C_R=C_{d.l}$ double-diffuse layer at an interface (microfarads/cm^2)
F: Film, resistive surface layer	R_F - true surface region resistance from film such as SiO_2 (or hydrate) on glass.	C_F - same as C_R
A: Adsorption, reaction, coupled cases, mainly observed for metals	R_A - surface resistance equivalent to adsorption for systems so irreversible that dissolved species do not react, but specifically adsorbed species have finite, potential-dependent rates.	$C_A=C_{d.l}$ doubly-diffuse capacitance modified by specific adsorption of charged and uncharged species.
G: Generation of charge carriers, recombination of carriers	R_G - modification of R_W by internal replacement or removal of charge carriers from lattice sites, ion pairs, or complexes.	C_G=pseudo capacitance (like C_W) because of natural charge separation in concentration profiles of species with different mobilities.
W: Warburg diffusion of charge carriers and complex species	R_W = resistance near inner membrane surface occurring by addition of charge carriers from electrolyte at one interface, and loss of charge carriers from opposite interface during halfcircle of current flow.	C_W=pseudo capacitance in mixtures of ions moving at different velocities because of different mobilities.

TABLE II

Impedance parameters from least-squares regression (16)

Type	$AgNO_3$ (M)	Ω (rps)	$\theta(\Omega)$	$C_{d1}(\mu F)$	$\bar{\sigma}(\Omega\,s^{-1/2})$	$\gamma(\Omega)$	$K(s^{-1})$
HP2	10^{-4}	36	93 ± 3	6.0 ± 0.2	406 ± 15	15 ± 10	24.1 ± 1.0
HP3	10^{-4}	36	44 ± 3	9.3 ± 1.1	502 ± 8	63 ± 2	5.1 ± 0.1
CP2	10^{-4}	36	24 ± 23	12.3 ± 19.1	426 ± 49	34 ± 5	2.1 ± 0.2
CP3	10^{-4}	36	25 ± 7	8.9 ± 6.5	406 ± 16	36 ± 3	3.3 ± 0.3
HP2	10^{-5}	36	150 ± 12	4.0 ± 4.7	1242 ± 66	662 ± 21	5.8 ± 0.3
HP3	10^{-5}	36	126 ± 39	5.4 ± 2.3	1185 ± 147	484 ± 45	5.2 ± 0.4
HP1	10^{-5}	36	30 ± 11	7.2 ± 1.4	1268 ± 50	379 ± 19	7.0 ± 0.4
CP3	10^{-5}	36	85 ± 38	6.3 ± 2.4	1515 ± 91	458 ± 22	2.9 ± 0.1
CP2	10^{-5}	20	149 ± 14	8.2 ± 0.7	1174 ± 61	967 ± 27	3.5 ± 0.1
CP2	10^{-5}	36	125 ± 16	6.7 ± 0.8	1283 ± 68	687 ± 22	5.1 ± 0.2
CP2	10^{-5}	49	108 ± 16	6.3 ± 0.9	1377 ± 66	571 ± 18	6.1 ± 0.2

TABLE III

Bulk Properties of Dry Aliquat Nitrate in Nitrobenzene (18)

Aliquat conc. (M)	Temp. (C°)	R_∞ (ohm-cm)	σ_0 (μSiemen/cm)	C_g (pf/cm)	K (C_g/C_g vac)	$\tau_\infty = \tau_B$ (μs)
0.0	20.0	60.2M	0.0166	3.23±.03	36.5±.3	194
	25.0	55.6	0.0178	3.19±.05	36.0±.5	177
	30.0	46.9	0.0213	3.13±.06	34.8±.7	147
	35.0	38.5	0.0260	2.94±.06	33.2±.7	113
1.6×10^{-5}	20.0	2.45M	0.409	3.16±.15	35.7±1.7	7.74
	25.0	2.29	0.437	3.08±.19	34.8±2.1	7.05
	30.0	2.14	0.467	2.96±.16	33.4±1.8	6.33
	35.0	2.02	0.495	2.87±.20	32.4±2.2	5.80
4.0×10^{-5}	20.0	0.922M	1.09	3.10±.13	35.0±1.4	2.86
	25.0	0.857	1.17	2.94±.08	33.2±0.9	2.52
	30.0	0.804	1.25	2.88±.10	32.5±1.3	2.32
	35.0	0.747	1.34	2.76±.10	31.1±1.1	2.05
8.0×10^{-5}	20.0	0.452M	2.21	3.08±.19	34.8±2.2	1.39
	25.0	0.419	2.39	2.96±.22	33.5±2.5	1.24
	30.0	0.392	2.55	2.82±.27	31.9±3.1	1.11
	35.0	0.371	2.67	2.79±.34	31.5±3.8	1.04
2.0×10^{-4}	20.0	0.184M	5.45	2.99±.19	33.8±2.1	0.550
	25.0	0.170	5.90	2.86±.04	32.3±0.5	0.486
	30.0	0.157	6.39	2.67±.15	30.1±1.7	0.419
	35.0	0.146	6.87	2.52±.11	28.5±1.3	0.368
4.0×10^{-4}	20.0	93.7K	10.7	2.40±.21	27.1±2.3	0.225
	25.0	87.1	11.5	2.26±.31	25.5±3.5	0.197
	30.0	81.4	12.3	2.12±.15	24.0±1.6	0.173
	35.0	76.5	13.1	1.98±.28	22.4±3.2	0.151
8.0×10^{-4}	20.0	48.3K	20.7			
	25.0	45.2	22.1			
	30.0	42.2	23.7			
	35.0	39.1	25.6			

TABLE III - continued

Aliquat conc. (M)	Temp. (C°)	R_∞ (ohm-cm)	σ_0 (μSiemen/cm)
2.0×10^{-3}	20.0	21.2K	42.2
	25.0	19.6	51.0
	30.0	18.5	54.0
	35.0	17.2	58.1
1.0×10^{-2}	20.0	5.52K	181
	25.0	5.15	194
	30.0	4.72	212
	35.0	4.35	230

TABLE IV

Activation Energy for Ionic Transport in Wet and Dry Exchangers (18)

Aliquat conc. (M)	E_a (kcal/mole)	
	Dry	Wet
1.6×10^{-5}	$2.26 \pm .04$	
4.0×10^{-5}	$2.47 \pm .11$	
8.0×10^{-5}	$2.41 \pm .05$	
2.0×10^{-4}	$2.77 \pm .04$	$2.52 \pm .08$
4.0×10^{-4}	$2.37 \pm .10$	$2.58 \pm .01$
8.0×10^{-4}	$2.62 \pm .08$	$2.38 \pm .03$
2.0×10^{-3}	$2.44 \pm .15$	$2.36 \pm .03$
1.0×10^{-2}	$2.91 \pm .11$	$2.32 \pm .10$
Mean value	$2.53 \pm .22$	$2.43 \pm .11$

TABLE V

K_A and Λ_0 Values for Aliquat Nitrate in Nitrobenzene (18)

State (Wet or Dry)	Temp. (°C)	K_A (M^{-1})	Λ_0
Dry	20.0	131	28.2
Dry	25.0	133	30.3
Dry	30.0	120	32.3
Dry	35.0	108	34.4
Wet	25.0	140	29.1

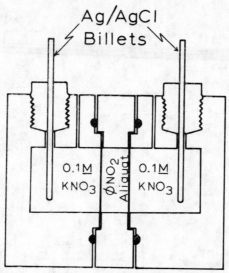

Fig. 1: Cross-sectional view of assembled, thick-membrane, reversible ion-contact impedance cell. The central support section for the liquid nitrobenzene-Aliquat exchanger can be replaced and single crystals or pressed pellet membranes inserted.

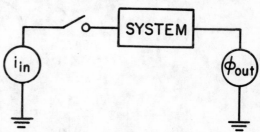

Fig. 2: "Thought" experiment circuit for step, pulse or periodic current perturbation of system.

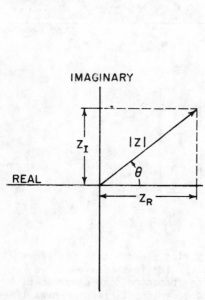

Fig. 3: Argand Diagram for Impedance Functions

$$Z = (Z_R^2 + Z_I^2)^{1/2}; \ Z_R; \ Z_I$$

and θ.

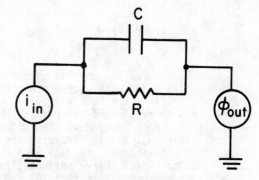

Fig. 4: Conventional impedance circuit arrangement for a parallel RC element.

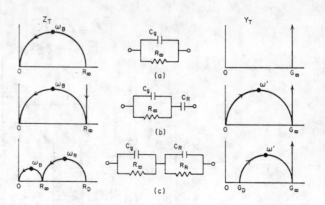

Fig. 5: Impedance and admittance plane plots and equivalent circuits for simple RC systems.

ω increases in the arrow direction.

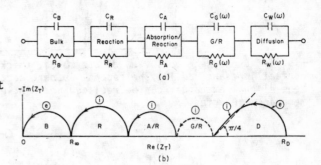

Fig. 6: (a) Equivalent circuit following from the recent models (Ref.6) (b) Possible $Z_T(\omega)$ complex plane plot consistent with the circuit in (a).

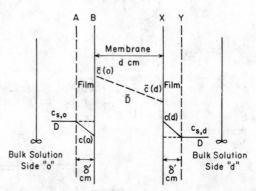

Fig. 7: Concentration profiles of cations during steady-state transport across a liquid ion exchange membrane under applications of a positive current. Solid lines are bulk and "Film" region concentrations under ideal diffusion with rapid surface equilibrium and moderate stirring. Broken lines in "Film" indicate ideal diffusion with rapid surface equilibrium for rapid external stirring. A,B,X,Y indicate the respective boundaries: bulk solution-film, film - membranes interface at X=O, membrane - film at X=d, and film - bulk solution. Space charge perturbations are omitted. At zero flux, the dashed internal profile is flat. δ' and δ'' are Nernst diffusion layer thicknesses. This drawing also applies in the absence of current to spontaneous flow profiles of neutral molecules and salts which have preferential solubilities in the membrane phase, moving from high (left side) concentration to low (right side).

42

Fig. 8: Potential response on transfer of an Ag_2S electrode from $AgNO_3$ 10^{-4} to 10^{-5} M. Points are experimental; curves are theoretical. $E°(exp.)=-296$ mV. Curve 1 best fit, using $i=1$, $E°=-292$ mV, $K_1=15$ mV, $\tau_1^{-1}=0.60$/min, best $E∞(calc.)=-277$ mV, $\sigma=1.1$ mV(SD). Curve 2 best fit, using $i=2$, $E°=-296$ mV, $k_1=10$ mV, $\tau_1^{-1}=5.0$/min, $k_2=9$ mV, $\tau_2^{-1}=0.24$/min, best $E∞(calc.)=-277$ mV, $\sigma=0.3$ mV(SD). General form: $E=E°+\Sigma k_i[1-exp(t/\tau_i)]$. (From Shatkay, A., Anal. Chem., 48, 1039, 1976. With permission.)

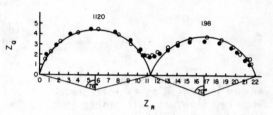

Fig. 9: Complex impedance plane representation of G.P. glass data at 45C. Dots are experimental points and circles are theoretical. Z_R and Z_I are given in $M\Omega$. ω at peak Z_I is written above each semicircle. Ref. 4.

Fig. 10a: In-phase G.P. glass impedance expressed as conductance G in ohm^{-1} frequency, f, in Hz. Data points (not shown) are spaced at quarter decade intervals. Ref. 4.

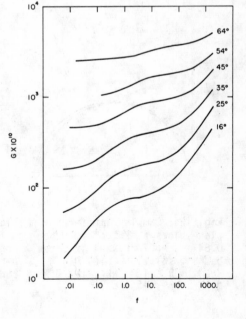

43

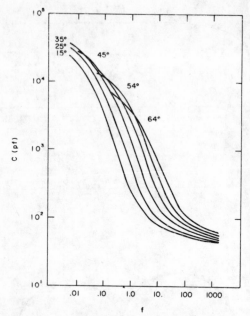

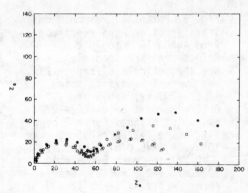

Fig. 10b: Out-of-phase G.P. glass impedance expressed as capacitance, C; frequency f in Hz. Data points (not shown) are spaced at quarter decade intervals. Ref. 4.

Fig. 11: Complex impedance plane representation of the results of the etching experiments. Z_R and Z_I are given in MΩ. (●) before etching (KCl solution contact); (□) before etching (Hg contact); (O) after 10% HF treatment (Hg contact), (△) after 10% HF and 18 M H_2SO_4 treatment (Hg contact). All measurements were made at 25C except the measurement denoted by (●) which was made at room temperature. Ref. 4.

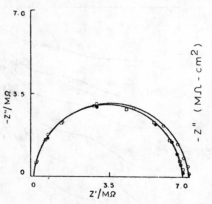

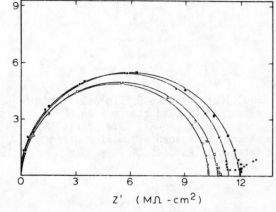

Fig. 12a: Complex impedance plane plots at 25C of polished 0.64 mm. AgCl crystal in three bathing solutions. (O) 0.1 M $AgNO_3$ (□) 0.1 M KNO_3, (●) 0.1 M KCL. Ref. 14.

Fig. 12b: Complex impedance plane plots for a 1.07 mm AgCl crystal. (□) first solution contact, (●) graphite DAG contact, (□) vapor deposited silver contact, (O) second solution contact. Ref. 14.

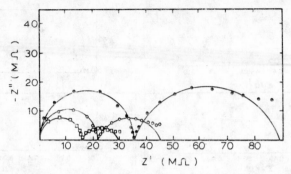

Fig. 13: Complex impedance plane plots for silver painted 4.00 mm AgCl crystal at three temperatures. (□) 35C, (0) 25C, (●) 15C. Ref. 14.

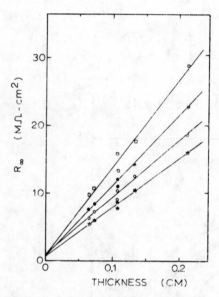

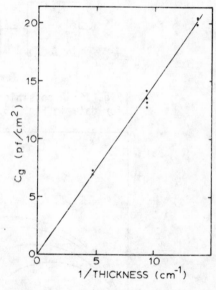

Fig. 14a: R_∞ vs. δ for etched AgCl crystals at all temperatures. (★) 35C, (0) 30C, (●) 25C, (□) 20C. Ref. 14.

Fig. 14b: C_g vs. $1/\delta$ for etched AgCl crystals at all temperatures. Ref. 14.

45

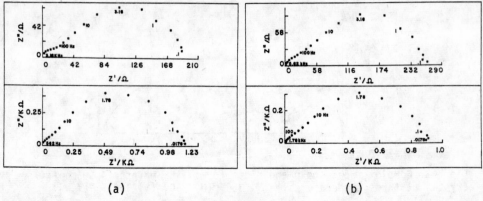

(a)   (b)

Fig. 15: Impedance plane plots at a rotation rate of 36 rps in (□) 10^{-4} M AgNO$_3$ (●) 10^{-5} M AgNO$_3$ for (a) cold-pressed Ag$_2$S membrane (excess S^{2-}) and (b) hot-pressed Ag$_2$S membrane (stoichiometric).   Ref. 16.

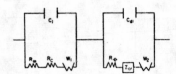

Fig. 16: A possible equivalent circuit for data in Fig. 15.

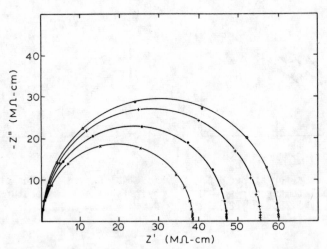

Fig. 17:   Impedance plane plot of Z_I vs. Z_R for purified nitrobenzene at four temperatures (0) 20C; (+) 25C; (●) 30C; (x) 35C.   Ref. 18.

Fig. 18a: Impedance plane diagram of 0.4 cm. thick AR grade nitrobenzene liquid membrane as affected by constraining polymer. Ref. 19.

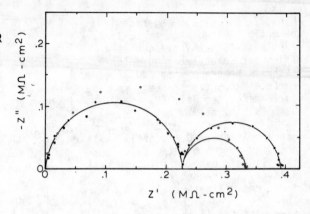

Fig. 18b: Impedance plane diagram of 0.4 cm. thick membrane of 1×10^{-5} M Aliquat nitrate in nitrobenzene. (O) platinum cell results; (●) reversible contact measurement with dialysis membrane constraint. Ref. 19.

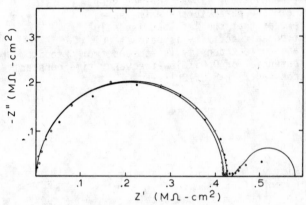

Fig. 19: Reversible ion-contact impedance plane plot for 0.01cm thick membrane of 8×10^{-4} M Aliquat nitrate in purified nitrobenzene in a 0.45µ PVC-acrylonitrile copolymer membrane.

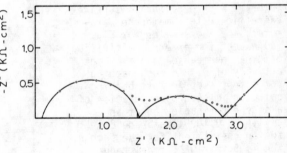

Fig. 20: Reversible contact impedance plane diagram of a cell as a function of time after fabrication. (●) new; (+) 24 hs; (O) 48 hs. Cell membrane system was 0.01 cm. thick, 0.45µ hole size, PVC-acrylonitrile containing 8×10^{-4} M Aliquat nitrate in nitrobenzene. Ref. 19.

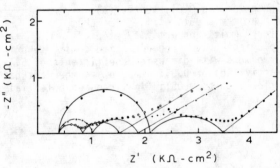

Fig. 21: Curve (a): logarithmic potential - time plot for a linear single-salt activity-step at a thick permselective liquid ion-exchange membrane with equilibrium surface extraction rates; membrane parameters (see Ref. 20) are $D_+ = D_- = 5.0 \times 10^{12}$, $d = 10^6$, $K = 5.0$, $K_{eq} = 1.0$, $k_b = 10^{15}$, and $I = 0$; membrane salt concentration is 0.1; initial salt activity in the bathing solutions is 0.1; left bathing solution salt activity is stepped to 0.101; final reduced potential is 0.01. Curve (b); logarithmic potential - time plot for linear current step; membrane parameters are the same, except that the left bathing solution salt activity is held at 0.1, and current density is stepped from 0 to 5000, resulting in a final reduced potential of 0.01; low-frequency time constant, $\tau_W = 2.0 \times 10^{-2}$; geometric time constant, $\tau_\infty = 5.0 \times 10^{-12} = \tau_B$.

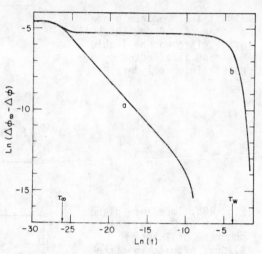

Fig. 22: Curve (a): logarithmic potential - time plot for a multi-ion activity step at a thick permselective liquid ion-exchange membrane with equilibrium surface extraction rates; membrane parameters are the same as for Fig. 21 (a) except for external activities; original permeable ion is at an activity of 0.1 in both bathing solutions; step in left bathing solution is to a pure solution of a second permeable ion at an activity of 0.1005; membrane diffusion coefficient for the second ion is 5.025×10^{12}; final reduced potential is 0.01. Curve (b): logarithmic potential - time plot for a multi-ion activity step displaying no low-frequency time constant; original bathing solutions contain two ions with diffusion coefficients of 5.1×10^{12} and 5.0×10^{12} at equal activities of 0.05; step raises the left bathing solution activity of the first ion to 0.051; final reduced potential is 0.01.

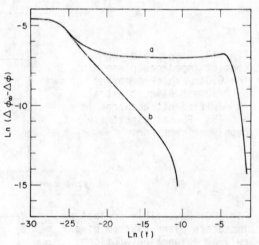

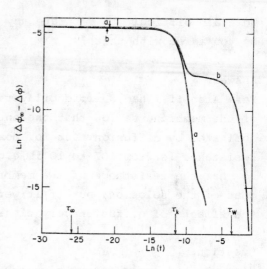

Fig. 23: Logarithmic potential – time plots for linear activity and current steps at a membrane with slow surface extraction rates. Curve (a) plots activity step; membrane parameters are given in Ref. 20. Curve (b) plots current step; membrane parameters are the same as for Fig. 21 (curve b), except that $k_b=1.0\times10^6$ and current density is stepped to 833. Final reduced potential is 0.01. $\tau_B = \tau_\infty = 5.0\times10^{-12}$, $\tau_k=1.0\times10^{-5}$, $\tau_W = 2.0\times10^{-2}$.

QUESTIONS AND COMMENTS

Participants of the discussion: R.P.Buck, G.Johansson, A.Lewenstam, E.Lindner, E.Pungor, J.D.R.Thomas, K.Tóth

Question:
I think you mentioned that you cannot deal problems of diffusion outside the membrane. Do you mean by "outside" at the membrane/solution interface or further afield ?

Answer:
The problem is that the small resistance due to the transport processes from the bulk to the interface cannot be discriminated from the total resistance. At very low frequencies, where the concentration change on the solution side is considerable, this measurement would be perhaps possible, but I think that the time scale of the concentration step measurements is more suitable for this type of study.

Comment:
I can add that if the resistance of the membrane is low, e.g. with sulphide membranes, the diffusion from the solution to the surface will dominate the measurements and it will

be again difficult to separate any surface reaction effect.
Do you agree with that ?

Answer:
Yes. Also if I had a lipid bilayer I could get the resistance
of the membrane so low that the entire experiment was
dominated by diffusion. But for most ISE-s that we have the
resistance is from 10^5 to $10^9 \Omega$ /e.g. in the case of glass/,
so that the resistance of the membrane always higher than
that of the solution, but if it were in reverse I would see
the effects of diffusion very clearly.

Question:
Does the a.c. measurement have any harmful effects on the
membranes ?

Answer:
I have never observed such effects but from other a.c. experiments I know that the a.c. method dissolves and rearranges the
surface. Our results were, however, unaffected by any eventual
rearrangement that may have occurred.

Comment:
I think that an eventual change in surface morphology could be
seen from response time measurements preceeding and following
the a.c. measurement.

Answer:
The morphology problem came after I have done most of the
work with the single crystals of the silver halides but the
calibration was never changed by the a.c. measurement. Since
the amplitude of the applied a.c. signal is quite low, e.g.
from 5mV to 1V, depending on the resistance of the membranes,
the total perturbation of the surface is not very great.

Question:
It seems to be difficult to measure the Warburg resistance
if you have a small circle on the plot. If we assume that it

could be correctly measured, what conclusions can you draw
from the measurement ?

Answer:
The Warburg impedance is measured at very low frequencies,
below say 0.01 Hz. Although we did not go down to the lowest
possible frequencies, others have done such measurements. The
Warburg impedance, as opposed to the high frequency resistance
which depends on all mobile ionic species in the membrane,
depends solely on those charge carriers which actually pass
through the interfaces.

Question:
How reproducible are your results ?

Answer:

Reproducibility depends first of all on the reproducibility
of the mounting of the membrane between the two rubber O-rings,
because this influences the available surface area. For this
reason all frequencies were applied without ever taking the
crystal out of the holder.
At high frequencies also the stray capacitance may cause some
error, so you need to bring the electrodes as well as the wires
very close to each other.

Question:
Where can you see the payoff of all these studies in improving
the quality of ion-selective electrodes ?

Answer:
Although we have learned some useful practical things about
the electrodes by these impedance studies e.g. that the surfaces of the single crystals need being properly polished, or
how proteins and other substances influence the electrode,
but I have not been able to predict e.g. how to make a good
phosphate electrode simply by having studied the impedances
of other electrodes.

Question:
How can you avoid measuring stray capacitance at high frequencies ?

Answer:
At high frequencies $\frac{1}{j\omega C}$ in the imaginary part becomes small, so that our plot is semicircular heading into the zero point. Stray capacitance causes scattering there, but since we know that we should get a circle therefore we can put the line through the points even though there is some scatter there. We could obtain better results for the value of capacitance in this region but we do not need them.
For any case we put our cell in a Faraday cage and therefore it is not effected by e.g. persons standing nearby it. But ghost effects are still there, so e.g. the dielectric constants of the crystals when calculated from our measurements are wrong by about 5-10 percent.

ELECTROLYSIS AT THE INTERFACE OF TWO IMMISCIBLE ELECTROLYTE SOLUTIONS AND ITS ANALYTICAL ASPECTS

J. KORYTA

J. Heyrovský Institute of Physical Chemistry and Electrochemistry,
Czechoslovak Academy of Sciences, Opletalova 25, 110 00 Prague 1, Czechoslovakia

ABSTRACT

Under certain conditions the interface of two immiscible electrolyte solutions (ITIES) has similar properties as the interface metallic electrode/electrolyte solution and can be investigated by the methods of electrochemical kinetics. The main applications of this phenomenon are as follows:

i. The selectivity and Nernstian, sub-Nernstian and super-Nernstian behaviour of liquid-membrane ion-selective electrodes can be assessed.

ii. The response-time of liquid-membrane ion-selective microelectrodes can be estimated.

iii. Several processes important in bioenergetics can be modelled by electrolysis at ITIES, namely carrier mediated transfer of protons and alkali-metal ions and redox-reaction driven proton transport across ITIES.

iv. A new biological membrane model, a phospholipid mono-layer adsorbed at ITIES, has been proposed.

v. Ion-carriers can be quantitatively determined by voltammetry at ITIES.

INTRODUCTION

The electrolysis at the interface of two immiscible electrolyte solutions (ITIES) has been introduced to the field of electroanalytical chemistry by Gavach and his coworkers (1,2) who used chronopotentiometry and in our laboratory (3-7) where the method of investigation was, besides chronopotentiometry, cyclic voltammetry and polarography with electrolyte dropping electrode. The main results have been described in several papers (8-12).

The two phases in contact are water and an organic liquid (the "oil" phase) of low mutual solubility. Organic solvents of appropriate properties are, for example, nitrobenzene(1,3) or 1,2-dichloroethane (7,13). When the aqueous solution contains a strongly hydrophilic electrolyte such as LiCl, $MgSO_4$ etc. and the oil phase a strongly hydrophobic electrolyte (consisting of cations such as tetrabutylammonium, Crystal Violet or tetraphenylarsonium and anions such as tetraphenylborate or dicarbollylcobaltate) ITIES shows behaviour similar to the interface metal electrode/electrolyte solution. There exists then a potential "window" where under polarization from an external source of electricity practically no electric current except the charging current flows across the ITIES.

SIMPLE ION TRANSFER

In voltammetric investigations the electrolytes mentioned in the preceding paragraph function as base electrolytes. The "edge" of the potential window is determined on the positive as well as on the negative side by the transfer of the ions of the base electrolyte. Thus, in the case of 0.05 M LiCl in the aqueous phase and 0.05 M tetrabutylammonium tetraphenylborate in nitrobenzene the potential window on the negative side is limited by the transfer of tetrabutylammonium cation from nitrobenzene to water while its positive boundary is fixed by the transfer of tetraphenylborate anion from nitrobenzene to water. The transfer of Li^+ from water to nitrobenzene would take place at more positive potentials than the transfer of tetraphenylborate anion from nitrobenzene to water while the transfer of Cl^- would take place at more negative potentials than the transfer of tetrabutylammonium cation in the opposite direction.

When semihydrophobic ions are present in one of the phases in contact they can be transferred from that phase to the other phase when an appropriate potential difference is imposed across ITIES. This can be achieved (5) by means of a four-electrode potentiostat (identical with the "voltage-clamp" method used in electrophysiology). This is the case of the simple ion transfer which has thoroughly been studied with cations such as those of tetraalkylammonium (1) and cesium (14) and with anions such as picrate (7), octoate and dodecylsulfate (15). Recently, the transfer of choline and acetylcholine has also been observed (14). The method which has most often been used in these investigations is cyclic voltammetry. The peak voltammograms obtained with this method perfectly obey the relationships deduced for reversible processes at mercury electrodes. The half-wave

potentials calculated from the voltammograms are electroche-
mically equivalent to the standard Gibbs energies of transfer
of each particular ion from the aqueous to the oil phase. At
rather high polarization rates the partial control of the
process by the rate of transfer could be detected. The basic
equation of the ion transfer (completely analogous to the
equation of the rate of electrode reaction) has the form

$$
j = z_i F k^0 \left\{ c_i(w) \exp\left[\alpha z_i F(\Delta_o^w \varphi - \Delta_o^w \varphi_i^0)/(RT) \right] \right.
$$
$$
\left. - c_i(o) \exp\left[-(1-\alpha) z_i F(\Delta_o^w \varphi - \Delta_o^w \varphi_i^0)/(RT) \right] \right\} \tag{1}
$$

where j is the current density, z_i the charge number of the
ion i, k^0 the standard rate constant of ion transfer, $c_i(w)$
and $c_i(o)$ are the concentration of the ion in the aqueous
phase and in the oil phase, respectively, α is the transfer
coefficient, $\Delta_o^w \varphi$ the electrical potential difference between the aqueous phase and the oil phase and $\Delta_o^w \varphi_i^0$ the standard potential difference between these phases for the ion i.
The remaining symbols have the usual significance.

However, the electrical potential in the electrical diffuse double-layer in both the aqueous and the oil phase con-
tributes strongly to the total potential difference between
these phases (17,18) so that the Frumkin correction for the
diffuse double-layer potential has always to be made (19).

In Table 1 the values of k^0 obtained without the Frumkin
correction have been listed. Obviously the simple ion transfer reactions are fast processes.

ELECTRON TRANSFER

The electron transfer across ITIES differs from the ion
transfer and from the electron transfer in the system metallic electrode/electrolyte solution (19a) in that the products
$c_{Red}(o)c_{ox}(w)$ and $c_{ox}(o)c_{Red}(w)$ enter the equation (1) in
place of $c_i(w)$ and $c_i(o)$. An example of this process is shown
in Fig.1. In the cyclic voltammetric mode the electron coming
from ferrocene in nitrobenzene reduces the ferricyanide ion
present in water on increasing the electrode potential while
the ferricinium ion is reduced by electrons from ferrocya-
nide in the reversed polarization course (20,21).

CATION TRANSFER FACILITATED BY IONOPHORES

In the presence of complex formers dissolved in the organic phase, which belong to the class of macrocyclic ionopho-

res like crown polyethers, macrotetrolide antibiotics or valinomycin the transfer of alkali metal cations from the aqueous into the organic phase is observed. Otherwise the standard potential of these ions (with the exception of cesium)
is so positive that tetraphenylborate is transferred before
these cations or both transfer processes coincide. When the
concentration of the alkali metal ions in the aqueous solution is considerably higher than that of the ionophore in
the organic solution the transfer of alkali metal ions is
characterized by reversible cyclic voltammograms (Fig.2).
The peak height is proportional to the concentration of the
complex former. The peak potential is independent of the
concentration of the ionophore and is shifted by 0.06 V to
more negative potential with a ten-times increase of the concentration of the alkali metal.

We explain this behaviour of alkali metal ions (M^+) in
the presence of the ionophore X by complexation in the organic phase,

$$M^+(o) + X(o) \rightleftharpoons MX^+(o) \tag{2}$$

accompanied with the transfer

$$M^+(w) \rightleftharpoons M^+(o) \tag{3}$$

In equilibrium we have

$$
\begin{aligned}
\Delta_o^w \varphi_{M^+} &= \Delta_o^w \varphi_{M^+}^o + (RT/F)\ln(a_{M^+}(o)/a_{M^+}(w)) \approx \\
&\quad \Delta_o^w \varphi_{M^+}^o + (RT/F)\ln\left\{\left[M^+(o)\right]/\left[M^+(w)\right]\right\} = \\
&= \Delta_o^w \varphi_{M^+}^o + (RT/F)\ln\frac{\left[MX^+(o)\right]}{\left[X(w)\right]K_{MX}\left[M^+(w)\right]}
\end{aligned}
\tag{4}
$$

where $\Delta_o^w \varphi_{M^+}^o$ is the standard potential of transfer of the
ion M^+ across the interface water/oil, $a_{M^+}(o)$ and $a_{M^+}(w)$ the
activities of the ion in water and in the oil phase, respectively and

$$K_{MX} = \frac{\left[MX^+(o)\right]}{\left[M^+(o)\right]\left[X(o)\right]} \tag{5}$$

is the stability constant of MX^+ in the organic phase.

Since $M^+(w)$ is in excess and its diffusion need not be considered the only diffusing species are $MX^+(o)$ and $X(o)$. For the polarographic half-wave potential we have the condition

$$D_{MX^+}^{1/2} \left[MX^+(o)\right]_O = D_X^{1/2} \left[X(o)\right]_O \qquad (6)$$

where D_{MX^+} and D_X are the diffusion coefficients of the complex and of the complex former in the organic phase and the index O indicates the concentrations close to the interface. The half-wave potential of the complex is then given by equation

$$\Delta\varphi_{1/2,MX^+} = \Delta_o^w\varphi_{M^+}^0 + (RT/F)\ln(D_X/D_{MX^+})^{1/2} - \qquad (7)$$
$$- (RT/F)\ln K_{MX} \left[M^+(w)\right] \; .$$

In view of this equation stronger complexation and higher concentration of the metal ion causes larger shift of the half-wave potential (and, in the same way, of the peak potential) to more negative potential.

It should be noted that in this way complexation of all alkali metal ions except lithium with dibenzo-18-crown-6 was observed. Thus both the concentration of the ionophore and the stability constant of its complex can be determined by means of cyclic voltammetry at ITIES (7,8).

The proton transfer facilitated by complexation with dibenzo-18-crown-6 has also been observed (22).

PROTON TRANSFER FACILITATED BY REDOX
REACTION IN THE OIL PHASE

The standard potential difference water/oil for proton transfer is rather positive so that no proton transfer is observed at voltammograms of an acidified aqueous base electrolyte solution. However, when in the oil phase both ferrocene and benzoquinone are present the proton transfer from the aqueous to the oil phase is observed (Fig.3). This phenomenon is due to the process (22)

$$2H^+(w) + Fc(o) + Q(o) \longrightarrow 2Fc^+(o) + H_2Q(o) \qquad (8)$$

When the aqueous phase is not acidified no reaction takes place.

The process described by equation (8) resembles the simul-

taneous proton and electron transfer which join the photosystem II and I in photosynthesis,

2e(photosystem II) + plastoquinone(m)
+ 2H$^+$(w) $\longrightarrow$ plastohydroquinone(m),

where (m) denotes the thylakoid membrane (cf.(23)).

PHOSPHOLIPID MONOLAYER AT ITIES

When dipalmitoyllecithin is dissolved in the oil phase
(nitrobenzene) a monolayer of the phospholipid is formed at
ITIES (24). This monolayer has a conspicuous effect on all
processes taking place at ITIES. The transfer of semihydrophobic ions (Fig.4) as well as the electron transfer and
transfer facilitated by ionophores (Fig.5) are inhibited.
Obviously, in order to get across the hydrocarbon part of the
phospholipid monolayer the ions have to be completely strip-
ped-off the hydration sheaths while when crossing ITIES
without the monolayer they can pass with a part of the hydration sheath.

This finding has an important consequence for the iono-
phore-facilitated ion transfer. When the cation had to be
complexed first before coming to the oil phase a kinetic limiting current had to be observed, particularly when this
process were inhibited by the monolayer. However, no influence of this kind has actually been observed. The originally
reversible process only turns irreversible which implies
that the ion transfer across the interface has been inhibited. Obviously, the cation first has to cross ITIES with
subsequent complexation by the ionophore.

In the presence of calcium and other alkaline earth metal
ions a striking increase of current is observed (Fig.6). At
constant potential the current increases during a definite
period of time to a saturation value. The magnitude of this
current excludes the possibility that it is linked to a
change of the capacity of the interface. In view of low concentration of lecithin in nitrobenzene this effect cannot be
caused by ion-pair formation in the organic phase. Thus,
most probably the ion-pair formation between calcium and dicarbollylcobaltate, which is the anion of the base electrolyte, can take place in the low permittivity region of the
long-chain alkyl groups of lecithin. The final concentration
of these ion-pairs in the low permittivity region can be
quite high, which would correspond to a high value of the
saturation current.

The electrolysis at ITIES can be utilized for assessment of the Nernstian response of ion-selective electrodes (25, 26).

In the presence of interferants the ion-selective electrodes behaves like a corroding metal electrode. Under these conditions the membrane potential is a complete analogy of the mixed potential as already pointed out by Cammann (27). The determination of the mixed potential is based on steady state polarization curves of various electrode processes taking place at a corroding electrode (see, for example (28)) These polarization curves are identical with steady state polarograms. In the case of an ion-selective electrode the useful polarization curves are either constructed using data, for example, from cyclic voltammetry at ITIES, or are directly measured by means of polarography with electrolyte dropping electrode. For determination of the necessary data the system with ITIES should possess the following properties: The organic phase should be similar to the membrane solvent, and the counter-ion in the base electrolyte should be similar to the ion-exchanging ion. The Nernstian response is determined from the intersections of the polarization curves with the zero-line of current. In the presence of interferants these intersections are shifted and the dependence of the new intersections on the logarithm of determinand concentration indicates the resulting response. An example of such analysis is shown in Figs.7 and 8. In Fig.7 the basic behaviour of the picrate electrode is shown. The base electrode of the organic phase contains 10 mM tetrabutylammonium picrate (TBA Pi), the aqueous electrolyte 0.5 mM (curve 1) and 1 mM (curve 1') Pi^-. The difference of intersection potentials ab is 18 mV which corresponds to the Nernstian slope. When the same counterion acted in a perchlorate electrode a sub-Nernstian performance would be observed (Fig.8). In a similar way a super-Nernstian behaviour of the picrate electrode in the presence of semihydrophobic cations could be predicted (25).

In the case of usual types of ion-selective electrodes the response-time depends on a number of factors, including the transport of the determinand to the surface of the membrane, the surface processes at the surface and, eventually, the transport in the membrane phase. In view of the microporous structure of the surface in the case of the solid and plastic-film based membranes this problem is often quite complicated. However, in the case of liquid-membrane ion-selective microelectrodes the problem of the transport in the aqueous phase can be completely removed in the immersion experiment (29) or completely controlled in the iontophoretic experiment (30). Thus, in the absence of an interferant we have to deal with the comparatively simple situation when the electrode having an initial membrane potential $\Delta \varphi_i$ de-

pending on an equilibrium concentration of the determinand
is brought into contact with another concentration of the determinand in the analysed solution c_s. The rate with which
the potential $\Delta\varphi$ shifts from the value of $\Delta\varphi_i$ to the resulting equilibrium value $\Delta\varphi_f$ depends on the rate of charging the electrical double-layer formed at the interface mem-
brane/aqueous solution and on the rate of transfer of the determinand ion across this interface.

The rate of potential change can be described by a simple
equation where the rate of charging the double-layer is put
equal to the rate of ion transfer across the interface

$$v_c = v_t \tag{9}$$

The quantity v_c is given by equation for the charging current

$$v_c = -C \frac{d\eta}{dt} \tag{10}$$

where C is the capacity of the interface and the overpoten-
tial η is given by the difference $\Delta\varphi - \Delta\varphi_f$. For the
transfer kinetics we use the basic equation of electrochemi-
cal kinetics (1) in the form (31)

$$v_t = j^o \left[\exp \frac{\alpha F \eta}{RT} - \exp \left(-\frac{(1-\alpha) F \eta}{RT} \right) \right] \tag{11}$$

where j^o (A cm^{-2}) is the exchange current density, α is the
charge transfer coefficient and F, R and T have the usual
significance. For small values of η (when $|\Delta\varphi_f - \Delta\varphi_i|$
$\ll \frac{RT}{F}$) equation (11) can be linearized,

$$v_t \approx j^o F \eta / (RT) \tag{12}$$

By combining (9), (10) and (12) we obtain a simple differential equation

$$-C (d\eta/dt) = j^o F \eta / (RT) \tag{13}$$

The response-time τ will be defined as the time interval
during which the overpotential η changes from its initial
value $\eta_i \equiv \Delta\varphi_i$ to the final value $\eta_f = 0.05 \eta_i$ (τ_{95}).
By solution of (13) we obtain

$$\tau = \frac{RT}{F} \cdot \frac{C}{j^0} \ln \frac{\eta_i}{\eta_f} = \frac{RT}{F^2} \cdot \frac{C}{k^0 c_m^\alpha c_s^{(1-\alpha)}} \ln \frac{\eta_i}{\eta_f} \quad , \quad (14)$$

$$\tau_{95} = \frac{3.0\ RT}{F^2} \cdot \frac{C}{k^0 c_m^\alpha c_s (1-\alpha)} \tag{15}$$

where $k^0 (c.s^{-1})$ is the standard rate constant of ion transfer and c_m and c_s the concentrations of the determinand in the membrane and in the outer solution $(mol.cm^{-3})$, respectively.

While for the ion transfer across the interface water/membrane solvent the values of k^0 are not known some information can be obtained from the values of these constants measured at the interface water/nitrobenzene which are listed in Table 1.

Thus, for 10 mM membrane electrolytes and mM determinands in the outer aqueous phase and for the values of C equal to 10 - 20 μF cm^{-2} the values of τ_{95} are of the order of milliseconds which is comparable with the results found with ion-selective microelectrodes (30,32).

CONCLUSION

The applications of the electrolysis at ITIES to electroanalytical chemistry (stability constant determination, determination of ionophore concentration, assessment of Nernstian behaviour and of response time of ion-selective electrodes) and to biophysics (modelling of membrane processes such as ionophore mediated ion transport and redox-reaction driven ion transport, phospholipid monolayer at ITIES as an analogy of bilayer lipid membrane) show good perspective to this new methodical approach.

REFERENCES

1. C.Gavach and F.Henry, J.Electroanal.Chem.Surface Electrochem. 54, 361 (1974)
2. C.Gavach and B.D'Epenoux, J.Electroanal.Chem.Surface Electrochem. 55, 59 (1974)
3. J.Koryta, P.Vaný sek and M.Březina, J.Electroanal.Chem. Surface Electrochem. 64, 263 (1976)
4. J.Koryta, P.Vaný sek and M.Březina, J.Electroanal.Chem. Surface Electrochem. 75, 211 (1977)
5. Z.Samec, V.Mareček, J.Koryta and W.Khalil, J.Electroanal. Chem.Surface Electrochem. 83, 393 (1977)
6. Z.Samec, V.Mareček, J.Weber and D.Homolka, J.Electroanal.

Chem.Surface Electrochem. 99, 385 (1979)
7. D.Homolka, Le Q.Hung, A.Hofmanová, M.W.Khalil, J.Koryta, V.Mareček, Z.Samec, S.K.Sen, P.Vanýsek, J.Weber, M.Březina, M.Janda and I.Stibor, Anal.Chem. 52, 1606 (1980)
8. J.Koryta, Electrochim.Acta 24, 293 (1979)
9. J.Koryta and P.Vanýsek, in.Advances in Electrochemistry and Electrochemical Engineering (H.Gerischer and C.W.Tobias ed.) Vol.13, Wiley-Interscience, New York, in press.
10. Z.Samec, V.Mareček, P.Vanýsek and J.Koryta, Chem.listy 74, 715 (1980)
11. J.Koryta, M.Březina, A.Hofmanová, D.Homolka, Le Q.Hung, M.W.Khalil, V.Mareček, Z.Samec, S.K.Sen, P.Vanýsek and J.Weber, Bioelectrochem.Bioenerget. 7, 61 (1980)
12. J.Koryta, Hungarian Sci.Instr., in press.
13. Z.Koczorowski and G.Geblewicz, J.Electroanal.Chem. 108, 117 (1980)
14. Z.Samec, V.Mareček and J.Weber, J.Electroanal.Chem. 100, 841 (1979)
15. P.Vanýsek, J.Electroanal.Chem., in press.
16. M.Behrens-Peter and P.Vanýsek, unpublished results.
17. M.Gros, S.Gromb and C.Gavach: J.Electroanal.Chem. 89, 29 (1978)
18. B.D'Epenoux, P.Seta, G.Amblard and C.Gavach, J.Electroanal.Chem. 99, 77 (1979)
19. Z.Samec and D.Hájková, unpublished results.
19a. Z.Samec, J.Electroanal.Chem. 99, 197 (1979)
20. Z.Samec, V.Mareček and J.Weber, J.Electroanal.Chem. 96, 245 (1979)
21. Z.Samec, V.Mareček and J.Weber, J.Electroanal.Chem. 103, 11 (1979)
22. A.Hofmanová, Science, in preparation.
23. H.T.Witt, Quart.Rev.Biophys. 4, 365 (1971)
24. J.Koryta, Le Q.Hung and A.Hofmanová, Nature, in preparation.
25. J.Koryta, Anal.Chim.Acta 111, 1 (1979)
26. J.Koryta, Mitteilungsblatt Chem.Ges.DDR, in press.
27. K.Cammann, Anal.Chem. 50, 936 (1978)
28. J.Koryta, J.Dvořák and V.Boháčková, Electrochemistry, Science Paperbacks, Chapman and Hall, London 1973, p.318.
29. F.Vyskočil and N.Kříž, Pflügers Arch. 337, 265 (1972)
30. H.D.Lux and E.Neher, Exp.BrainRes. 17, 190 (1973)
31. K.J.Vetter, Elektrochemische Kinetik, Springer-Verlag, Berlin 1961, p.122.
32. E.Ujec, private communication.

Table 1. Parameters of the charge transfer reaction (2,14)

Ion	α	$k^o/(cm.s^{-1})$
tetraethylammonium	0.35	2.2×10^{-3}
tetrapropylammonium	0.27	2.3×10^{-3}
tetrabutylammonium	0.23	5×10^{-3}
Cs^+	0.46	5.5×10^{-2}

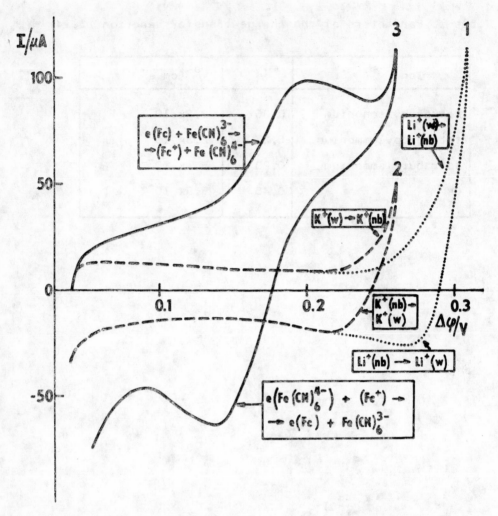

Figure 1. Electron transfer across ITIES: 1 - w : 0.05 M LiCl; nb : 0.1 M ferrocene, 0.05 M tetrabutyl-ammonium tetraphenylborate; 2 - w : 1 mM $K_3 Fe(CN)_6$, 1 mM $K_4 Fe(CN)_6$, 0.05 M LiCl; nb : 0.05 M tetra-butylammonium tetraphenylborate; 3 - w : 1 mM $K_3 Fe(CN)_6$, 1 mM $K_4 Fe(CN)_6$, 0.05 LiCl; nb : 0.1M ferrocene, 0.05 M tetrabutylammonium tetraphenyl-borate; upper curve : reduction of $Fe (CN)_6^{3-}$ (w = water) by electrons from ferrocene (nb = nit-robenzene); lower curve : reduction of ferricinium ion (nb) by electrons from $Fe (CN)_6^{4-}$ (w).

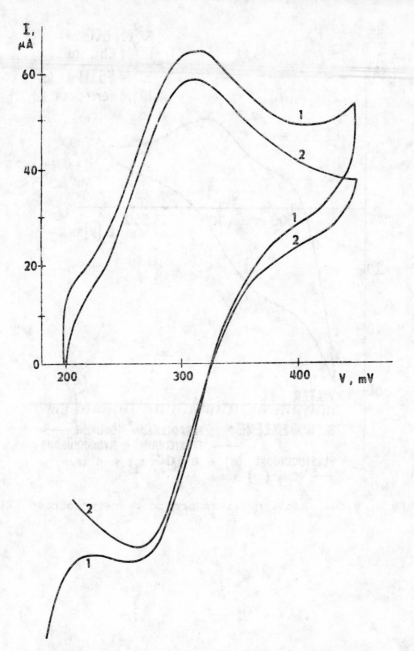

Figure 2. Cyclic voltammogram of 0.75 mM macrotetrolide anti-
biotic nonactin dissolved in nitrobenzene. Base
electrolytes: 0.082 M NaCl (aqueous phase), 0.05 M
tetrabutylammonium tetraphenylborate (nitrobenzene).
Voltage-scan rate 50 mV s^{-1}. 1 – original curve,
2 – after subtraction of the current of the base
electrolytes.

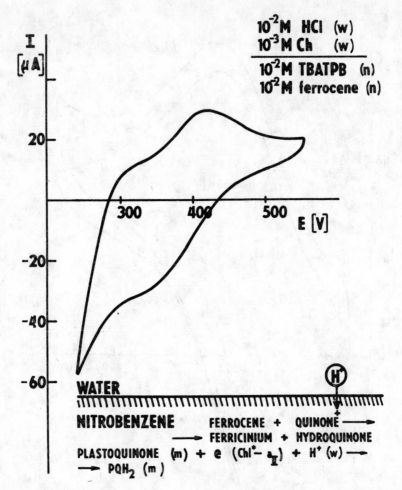

Figure 3. Redox reaction driven proton transfer across ITIES.

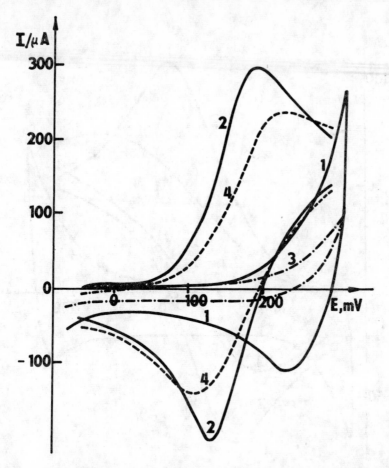

Figure 4. The influence of lecithin monolayer on voltammetric
curves of Cs⁺ at 5°C. 1 - base electrolytes, 10 mM
LiCl(w) and 17 mM tetrabutylammonium dicarbollylcobaltate (nitrobenzene), 2 - base electrolytes +
0.5 mM CsCl(w), 3 - base electrolytes + 0.17 mM
lecithin (nitrobenzene), 4 - base electrolytes +
0.5 mM CsCl(w) + 0.17 mM lecithin (nitrobenzene).
Polarization rate 100 mV/s.

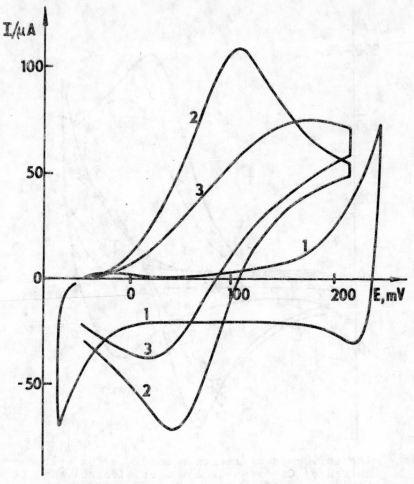

Figure 5. The influence of lecithin monolayer on voltammetric curves of Na^+-transfer facilitated by dibenzo-18-crown-6. 1 - 10 mM NaCl(w), 17 mM tetrabutylammonium dicarbollylcobaltate (nitrobenzene), 2 - after addition of 0.5 mM dibenzo-18-crown-6 (nitrobenzene) 3 - after addition of 0.15 mM lecithin (nitrobenzene) to the system 2. 5^oC, polarization rate 100 mV/s.

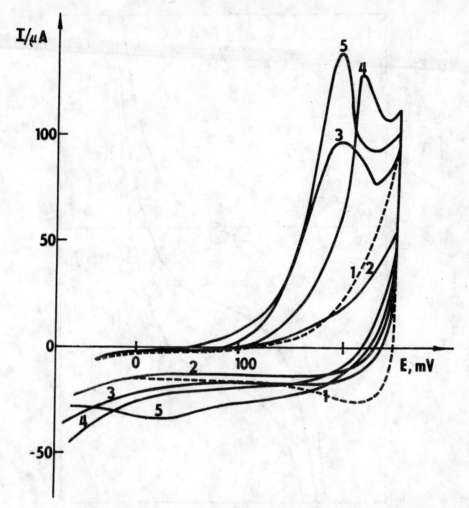

Figure 6. The influence of divalent metal cations on lecithin monolayer. 1 - base electrolytes, 10 mM LiCl(w) and 17 mM tetrabutylammonium dicarbollylcobaltate (nitrobenzene). 2 - 0.15 mM lecithin (nitrobenzene) added. 3 - after addition of 2 mM $CaCl_2$(w) to the system 2, 4 - after addition of 2 mM $BaCl_2$ to the system 2, 5 - after addition of 2 mM $MgCl_2$ to the system 2. 20°C, polarization rate 50 mV/s.

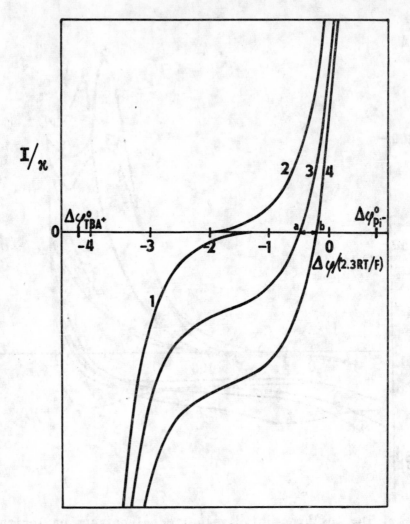

Figure 7. Polarograms calculated for: (1) 10 mM tetrabutylammonium cation in the organic phase: (2) 10 mM
picrate in the organic phase; (3) 0.5 mM picrate in
the aqueous phase and 10 mM tetrabutylammonium picrate in the organic phase; (4) 1 mM picrate in the
aqueous phase and 10 mM tetrabutylammonium picrate
in the organic phase. The shift of the zero-current
potentials ab is 2.3 RT/F.log 2 = 18 mV (Nernstian
slope). Equal diffusion coefficients of picrate in
the aqueous and organic phase are assumed.

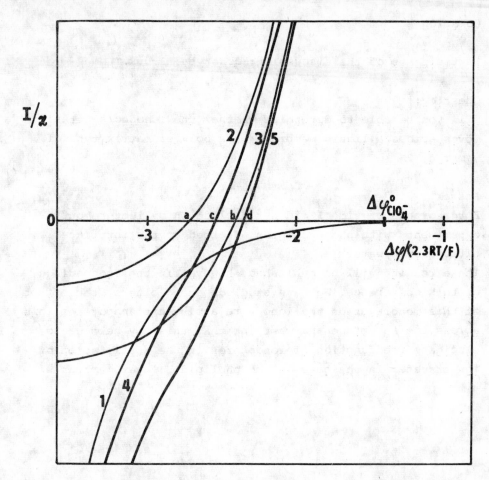

Figure 8. Polarograms calculated for: (1) 10 mM tetrabutylammonium cation in the organic phase; (2) 0.5 mM
ClO_4^- in the aqueous phase and 10 mM ClO_4^- in the organic phase; (3) 1 mM ClO_4^- in the aqueous phase and
10 mM ClO_4^- in the organic phase. By addition of
curves 1 and 2 and curves 1 and 4, composite polarograms 4 and 5 are obtained which correspond to
the voltammetric behaviour of the picrate electrode
with 0.5 mM picrate and 1 mM picrate, respectively,
in the aqueous phase. Without the influence of the
exchanging cation (tetrabutylammonium cation), the
shift of the zero-current potentials would be Nernstian (ab = 2.3 RT/F = 18 mV) whereas under the influence of tetrabutylammonium cation a sub-Nernstian slope (cd = 12 mV) is observed.

QUESTION

Participants of the discussion: J.Koryta, J.D.R.Thomas

Question:
Will you be able to determine whether the ionophore actually
moves across a liquid membrane or a polymer matrix type of
membrane ?

Answer:
For shortage of time I did not include in my lecture our
experiments with phospholipids adsorbed at the interface
nitrobenzene/water.
These results will be published, however in the Proceedings
of this conference. We have studied, among others, the effect
of this monolayer on the ionophore assisted transport of ions
across the interface and we think that this may help us to
elucidate the function of ionophores in the facilitation of
ion transfer in the case of phospholipid bilayer membranes.

TRANSPORT OF IONS THROUGH NEUTRAL CARRIER MEMBRANES

P. OGGENFUSS, W. E. MORF, R. J. FUNCK, H. V. PHAM, R. E. ZÜND,
E. PRETSCH AND W. SIMON

Swiss Federal Institute of Technology, Department of Organic Chemistry,
CH-8092 Zürich, Switzerland

ABSTRACT

Electrodialytic transport of ions has been studied on solvent polymeric membranes that contained neutral carriers selective for Li^+, Na^+, K^+, Ca^{2+}, Ba^{2+}, Cd^{2+}, and other ions,
respectively. Throughout, the measured transport numbers
attest a high selectivity in ion permeation. Specific cation
transport has been realized even under zero-current conditions.
A selective Ca^{2+}-pump, driven by a transmembrane pH gradient,
was designed by adding proton carriers to a calcium-carrier
membrane. The role of neutral carriers in anion permeation is
also discussed. Certain ethanolamine derivatives and tin-
organic compounds seem to facilitate the permeation of bicarbonate ions and have been introduced in liquid membrane electrodes selective for HCO_3^-.

INTRODUCTION

Neutral carriers have been widely used as the active components of cation-selective liquid membrane electrodes. So far,
ligands specific for Li^+, Na^+, K^+, NH_4^+, Ca^{2+}, Ba^{2+}, Cd^{2+},
and other ions have been introduced (see Fig. 1) and have
found acceptance in ion-sensor applications [1-5].

In a series of earlier contributions [2-4, 6-10] we have
shown that there exists a close relationship between the potentiometric ion selectivity of liquid membrane electrodes and
the selectivity exhibited by the same membranes in ion transport experiments. Two different types of transport studies can
be performed. In electrodialysis experiments [4, 6-8] a voltage is applied across the bulk membrane and the transport
number of the permeating ion is determined (i. e. the fraction
of electric current carried by this species). In zero-current
countertransport systems [9-12], on the other hand, a selective transport of cations across the membrane is induced and

compensated by an oppositely directed flow of a second sort of cations.

Here we summarize some recent results on ion transport through neutral carrier membranes.

ELECTRODIALYTIC CATION TRANSPORT

The striking ion selectivity of neutral-carrier-based membranes can be understood as a permeability selectivity. Thus, the same membranes, as are used in potentiometric sensors for specific ions, should be able to selectively pump these ions when a transmembrane potential is applied. In the case of ideally specific systems, the electric current should be carried exclusively by one sort of ions. This corresponds to a transport number of $t_i = 1$ for the permeating species.

Such ideal selectivity in ion transport is generally observed for neutral carrier membranes. A selection of experimental data is given in the Table. It includes some representative values from earlier measurements [8] as well as new results. Evidently, transport numbers close to unity have been realized for virtually all carrier membranes. The only exception is a cadmium-selective membrane exposed to a $CdCl_2$ solution where an apparent transference number of around 2 (referring to the divalent cation) was obtained. In this system, however, association between cationic complexes and anions is believed to occur and, hence, it is the single-charged ion pair $CdCl^+$ that probably is the permeating species. This hypothesis is corroborated by the e.m.f. response of this membrane system which is characteristic of a single-charged cation state [4]. Similar e.m.f. behavior was found for an uranyl-selective membrane [5].

Measurements on Na^+-, K^+-, and Ca^{2+}-selective membranes were also performed in the presence of possible interferents (see Table). While sodium-carrier membranes are evidently subject to some interference by potassium ions, the latter two systems successfully reject any ions other than the primary ion. Accordingly, neutral carrier membranes may be very attractive for electrolytic separations of ions.

ZERO-CURRENT COUNTERTRANSPORT OF CATIONS

Cation transport under zero-current conditions was first realized for liquid membranes with negatively charged ligands [9, 11, 12]. The primary energy source for such transport of Na^+ [11, 12] or Mg^{2+} ions [9] came from the simultaneous counterflow of hydrogen ions being driven by a pH gradient. Application of the same ion-pumping principle to neutral-carrier-based membranes was problematic, however, since neutral ligands - in contrast to the negatively charged counterparts - hardly interact with protons. To overcome this, a proton carrier (a lipophilic weak acid) was incorporated into the membrane, in addition to the neutral ionophore [10]. This finally allowed us to couple the net transport of specific cations to the facilitated diffusion of hydrogen ions. A detailed description of this counter-transport system as well as a theoretical treatment are given elsewhere [10].

Zero-current transport experiments were carried out on solvent polymeric membranes containing the calcium-selective ligand 7 (Fig. 1) and various proton carriers [10]. It was clearly shown that the system is capable of pumping ions only in the presence of proton carriers. The highest transport rates were achieved with the components OCPH and FCCP [10], both of which are highly lipophilic acids of $pK_A \approx 6$ that are commonly used as uncouplers of oxidative phosphorylation in mitochondria [13, 14]. The pronounced ion specificity of this "ion pump" is demonstrated in Fig. 2. Evidently, the membrane was able to pump a considerable number of Ca^{2+} ions whereas virtually no transport could be realized for Ba^{2+}, Mg^{2+}, and Na^+ ions. Attempts at transporting K^+ across the same membrane were also unsuccessful. The selectivity order documented in Fig. 2,

$$Ca^{2+} >> Ba^{2+} > Mg^{2+} \quad \text{and} \quad Ca^{2+} >> Na^+ \text{ (or } K^+\text{)},$$

agrees with potentiometric selectivity data reported for membranes based on ligand 7 ($K_{CaBa}^{Pot} < 10^{-3}$ and $K_{CaM}^{Pot} \approx 10^{-5}$ for $M^{z+} = Mg^{2+}$, Na^+, or K^+ [1-3]).

These results demonstrate that highly selective or even specific ion transports can be performed on neutral carrier membranes. It is conceivable that such ion-pumping systems could be exploited for selective ion separations.

ANION PERMEATION

Recently it was claimed that the ligands 1 and 2 (Fig. 1) may behave as ionophores for anions in bilayer lipid membranes [15]. In bulk membranes, however, only cation-binding properties are observed for these and related ligands although anions may be co-transported when forming ion pairs with the cationic complexes. Nevertheless, there is no reason to deny the existence of molecules that could serve as neutral carriers for anions in lipophilic membranes. Potential candidates for anion-selective ionophores were sought, so far, among two classes of compounds, namely ethanolamine derivatives and tin-organic compounds.

Ephedrine is known to form distinct complexes with phosphate ions [16]. This fact induced us to design and synthesize the highly lipophilic ligand N-octadecyl-ephedrine [17]. Liquid membranes consisting of 1.0 wt.-% ligand, 1.2 wt.-% potassium p-chloro-tetraphenylborate, 65.2 wt.-% butanoic acid 10-hydroxy-decyl ester, and 32.6 wt.-% PVC were used in electrode cells of the following type:

Hg; Hg_2Cl_2; KCl (satd.) | 1 M Li-acetate | sample solution
(buffered to pH 8)

| membrane | 0.01 M NaCl ; AgCl; Ag.

Such sensors did not respond to phosphate ions but showed rather high selectivity for bicarbonate ions (see Fig. 3). A selectivity factor $K_{HCO_3,Cl}^{Pot}$ of $1.8 \cdot 10^{-2}$ was determined by the fixed interference method (0.1 M solution of chloride [17])

The response mechanism of this bicarbonate-selective electrode has not been fully elucidated as yet. It was shown by ^{13}C-n.m.r. studies on the same type of membrane that nearly 100% of the ligand molecules are present as non-protonated amines (at pH-values of the sample solution above ~3, see Fig. 4). Thus, the substituted ethanolamines clearly behave as neutral carriers that apparently induce or facilitate the overall permeation of bicarbonate ions across the membrane. There still remain two possible mechanisms of bicarbonate permeation:

1) The ligand (denoted by L) actually carries bicarbonate ions:

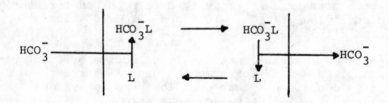

2) The ligand simply carries protons while carbon dioxide diffuses across the membrane:

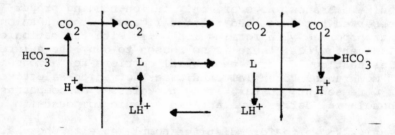

In practice, differentiation between the two mechanisms is not easy. Experimental work elucidating the exact pathway of bicarbonate permeation is in full progress [18].

An anion-carrier mechanism is also conceivable for membranes based on tin-organic compounds. Tripropyltin chloride was reported to induce chloride transport across erythrocyte membranes [19]. However, the incorporation of similar compounds into solvent polymeric membranes again resulted in bicarbonate-selective electrodes [20], the specifications being comparable to those given above.

ACKNOWLEDGEMENT

This work was partly supported by the Swiss National Science Foundation.

REFERENCES

[1]   W. E. Morf and W. Simon, in Ion-Selective Electrodes in Analytical Chemistry (H. Freiser, ed.), Plenum Press, New York, 1978.

[2]   W. E. Morf, The Principles of Ion-Selective Electrodes and of Membrane Transport, Akadémiai Kiadó, Budapest, 1980.

[3]   W. E. Morf, D. Ammann, R. Bissig, E. Pretsch, and W. Simon, in Progress in Macrocyclic Chemistry (R. M. Izatt and J. J. Christensen, eds.), Vol. 1, Wiley-Interscience, New York, 1979.

[4]   J. K. Schneider, P. Hofstetter, E. Pretsch, D. Ammann, and W. Simon, Helv. Chim. Acta 63, 217 (1980).

[5]   J. Šenkyr, D. Ammann, P. C. Meier, W. E. Morf, E. Pretsch, and W. Simon, Anal. Chem. 51, 786 (1979).

[6]   W. E. Morf, P. Wuhrmann, and W. Simon, Anal. Chem. 48, 1031 (1976).

[7]   W. E. Morf and W. Simon, in Ion-Selective Electrodes (E. Pungor and I. Buzás, eds.), Akadémiai Kiadó, Budapest, 1977, p. 25.

[8]   A. P. Thoma, A. Viviani-Nauer, S. Arvanitis, W. E. Morf, and W. Simon, Anal. Chem. 49, 1567 (1977).

[9]   D. Erne, W. E. Morf, S. Arvanitis, Z. Cimerman, D. Ammann, and W. Simon, Helv. Chim. Acta 62, 994 (1979).

[10] W. E. Morf, S. Arvanitis, and W. Simon, Chimia (Switzerland) 33, 452 (1979).

[11] E. L. Cussler, AIChE J. 17, 1300 (1971); E. L. Cussler, D. F. Evans, and M. A. Matesich, Science 172, 377 (1971).

[12] E. M. Choy, D. F. Evans, and E. L. Cussler, J. Am. Chem. Soc. 96, 7085 (1974).

[13] E. A. Liberman and V. P. Topaly, Biochim. Biophys. Acta 163, 125 (1968).

[14] O. H. LeBlanc, Jr., J. Membr. Biol. 4, 227 (1971).

[15] K.-H. Kuo and G. Eisenman, Biophys. J. 17, 212a (1977); R. Margalit and G. Eisenman, Arzneimittelforschung 28(I), 707 (1978).

[16] R. A. Hearn and C. E. Bugg, Acta Cryst. B28, 3662 (1972);
      R. A. Hearn, G. R. Freeman, and C. E. Bugg, J. Am. Chem.
      Soc. 95, 7150 (1973).
[17] W. Simon, D. Ammann, R. A. Dörig, D. Erne, R. J. J. Funck,
      H.-B. Jenny, E. Pretsch, and R. A. Steiner, in preparation.
[18] R. J. J. Funck, P. Schulthess, H. V. Pham, R. E. Zünd,
      W. E. Morf, D. Ammann, E. Pretsch, and W. Simon in prepa-
      ration.
[19] R. Motais, J. L. Cousin, and F. Sola, Biochim. Biophys.
      Acta 467, 357 (1977).
[20] R. E. Zünd, K. Hartman, R. J. J. Funck, and W. Simon, in
      preparation.

Table. Transport Numbers of Cation-Permselective Neutral Carrier Membranes

Cation studied	Membrane composition			Electrolytes		Transport number for cations studied[a] (95% confidence limits)
	Ligand; wt.-%	Solvent[b]; wt.-%	Matrix[c]; wt.-%	Anode compartment	Cathode compartment	
Li^+	1; 5.8	TEHP; 62.8	PVC; 31.4	10^{-3}M LiCl	10^{-3}M KCl	1.02 ± 0.21
Na^+	2; 3	DBS ; 65	PVC; 32	10^{-3}M NaCl	10^{-3}M KCl	0.92 ± 0.08
Na^+	3; 1[d]	DOS ; 66	PVC; 33	10^{-3}M NaCl	10^{-3}M KCl	0.94 ± 0.05
Na^+	3; 1	DOS ; 66	PVC; 33	10^{-3}M NaCl	10^{-3}M KCl	0.94 ± 0.19
Na^+	3; 1	DOS ; 66	PVC; 33	$5 \cdot 10^{-4}$M NaCl $5 \cdot 10^{-4}$M KCl	10^{-3}M KCl	0.81 ± 0.19
Na^+	4; 1	DOS ; 66	PVC; 33	10^{-3}M NaCl	10^{-3}M KCl	0.92 ± 0.05
Na^+	4; 1	DOS ; 66	PVC; 33	$5 \cdot 10^{-4}$M NaCl $5 \cdot 10^{-4}$M KCl	10^{-3}M KCl	0.64 ± 0.07
K^+	5; 3	DPP ; 67	PVC; 30	10^{-2}M KCl	10^{-2}M HCl	1.08 ± 0.07
K^+	5; 5	–	SR; 95	10^{-2}M KCl	10^{-2}M HCl	1.1 ± 0.15
K^+	5; 5	–	SR; 95	10^{-2}M KCl	10^{-2}M $HClO_4$	1.1 ± 0.15
K^{+e}	5; 1	DOA ; 66	PVC; 33	$9 \cdot 10^{-4}$M KCl	$9 \cdot 10^{-4}$M KCl	1.02 ± 0.04
$PEAH^{+f}$	6; 1	DOA ; 65	PVC; 34	$4 \cdot 10^{-3}$M PEAHCl	$4 \cdot 10^{-3}$M PEAHCl	0.95 ± 0.04
Ca^{2+}	8; 3	o-NPOE; 65	PVC; 32	10^{-3}M $CaCl_2$	10^{-3}M KCl	0.99 ± 0.08
Ca^{2+}	8; 3	o-NPOE; 65	PVC; 32	$5 \cdot 10^{-4}$M $CaCl_2$ $5 \cdot 10^{-4}$M $MgCl_2$	10^{-3}M KCl	0.99 ± 0.08
Ca^{2+}	8; 3	o-NPOE; 65	PVC; 32	$5 \cdot 10^{-4}$M $CaCl_2$ $5 \cdot 10^{-4}$M NaCl	10^{-3}M KCl	0.99 ± 0.02
Ca^{2+}	8; 3	o-NPOE; 65	PVC; 32	10^{-4}M $CaCl_2$	10^{-4}M KSCN	0.99 ± 0.02
Ba^{2+}	9; 1.1	o-NPOE; 65.9	PVC; 33	10^{-3}M $BaCl_2$	10^{-3}M KCl	0.99 ± 0.06
Cd^{2+}	10; 1	BHDE; 65	PVC; 34	10^{-2}M $Cd(NO_3)_2$	10^{-2}M $NaNO_3$	1.00 ± 0.03 1.01 ± 0.09
Cd^{2+}	10; 1	BHDE; 65	PVC; 34	10^{-2}M $CdCl_2$	10^{-2}M NaCl	1.95 ± 0.06[g] 2.01 ± 0.05[g]

[a] The values for Li^+, K^+, $PEAH^+$, and Ca^{2+} are taken from [8], and the data for Cd^{2+} are found in [4]. [b] TEHP: tris-(2-ethyl-hexyl)-phosphate; DBS: dibutyl sebacate; DOS: dioctyl sebacate; DPP: dipentyl phthalate; DOA: dioctyl adipate; o-NPOE: o-nitrophenyl octyl ether; BHDE: butanoic acid 10-hydroxy-decyl ester. [c] PVC: poly(vinylchloride); SR: silicone rubber. [d] The membrane contained ~0.3 wt.-% of the additive KSCN. [e] Radiochemical determination using the isotopes $^{42}K^+$ and $^{36}Cl^-$; the transport number for the anion was 0.0004±0.0002 [8]. [f] $PEAH^+$: ^{14}C-α-phenylethylammonium cation. [g] A transport number of ~1 is derived for the anion, which indicates a co-transport of chloride from the anode to the cathode compartment.

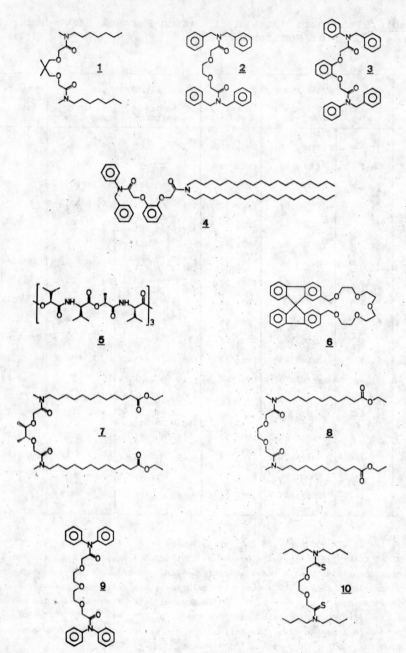

Figure 1. Structures of the neutral carriers used in ion transport studies on bulk membranes. The carriers are selective for
Li^+ ($\underline{1}$), Na^+ ($\underline{2}$-$\underline{4}$), K^+ ($\underline{5}$), phenylethylammonium ion ($\underline{6}$), Ca^{2+}
($\underline{7}$, $\underline{8}$), Ba^{2+} ($\underline{9}$), and Cd^{2+} ($\underline{10}$), respectively.

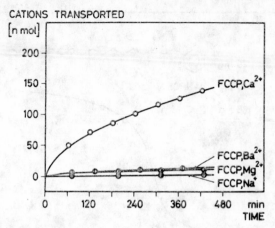

Figure 2. Selective cation transport induced by a pH gradient [10]. The membrane, containing the Ca^{2+}-selective ligand $\underline{7}$ and the proton carrier FCCP, was interposed between a $10^{-2}M$ solution (pH = 10.5) of a metal chloride MCl_z (M^{2+} = Ca^{2+}, Ba^{2+}, Mg^{2+}, Na^+) and a $10^{-2}M$ $KCl/10^{-3}M$ HCl solution.

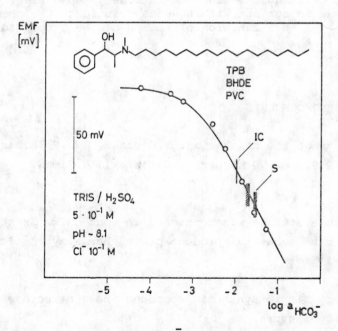

Figure 3. EMF-response of a HCO_3^--selective liquid membrane electrode based on the ligand N-octadecyl-ephedrine [17]. The bicarbonate samples were buffered to pH ~8.1 and contained 0.1M chloride. The physiological ranges of HCO_3^- activity in blood serum and intracellular fluid are indicated by S and IC, respectively.

81

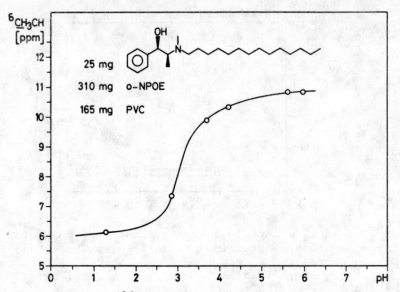

Figure 4. Results of ^{13}C-n.m.r. studies at 22.6 MHz on a PVC-
membrane containing the ligand N-tetradecyl-ephedrine. The
membrane was in contact with aqueous solutions. The chemical
shift is shown as a function of the pH of the sample solution
[18].

QUESTIONS AND COMMENTS

Participants of the discussion: Y.Umezawa, J.D.R.Thomas
K.Burger, R.P.Buck, G.Werner, A.Hulanicki, W.Simon

Question:
A magnesium electrode was mentioned in your talk. Is the
ionophore used a natural product other than a synthetic one ?

Answer:
It is a very simple synthetic compound whose structure was
shown in the lecture.

Question:
The diffusion coefficient in the membrane phase was found in
the order of 10^{-7} cm^2 sec^{-1}. Could you give us some indication
of how it was measured ?

Answer:
The determination was quite complicated. Basically, the
membrane was equilibrated with the solution and the amount of
the carrier in the solution was determined. This is one way,
and the other one is to use tracer technique. The details of
the measurement and of the mathematical approach are given
in the April issue of Analytical Chemistry. Most of the experimental and theoretical work has been carried out by Dr.E.
Pretsch.

Question:
It was mentioned that a small portion of the chloride ions
enter the membrane, and the electrical balance was also spoken
about. Would it be possible to give an explanation whether the
chloride ions gather at the outer side of the membrane ?

Answer:
The question is about the relative concentrations of anions
and cations in a neutral carrier membrane. The concentrations
were roughly 10 mmol of the carrier valinomycin, 0.4 mmol of
potassium ion and 0.01 mmol of chloride ion, which means about
forty times less chloride. The question is whether there is some
accumulation of chloride ions at the interface. There was no
indication of that and we have very strong evidence that there
was no kinetic limitation to the rather poor uptake of chloride
compared to potassium. It is simply a diffusion phenomenon.

Question:
Your suggestion was that you have hydroxide ions in the membrane
phase. I think that if you analyze the solution at equilibrium,
you can determine the hydroxide which is simply the result of
a deprotonation process. Have you done such an experiment ?

Answer:
Yes, of course we have done it. If we take a fresh membrane
which has not yet been conditioned in potassium chloride
solution, and immerse it into an aqueous solution containing
potassium chloride, then the solution gets more acidic,

indicating that protons are coming out of the membrane and entering the solution.

Question:
Did you get the equivalent amount ?

Answer:
Yes, roughly, as long as we could determine it with simple pH measurement. There was found no drastic unbalance.

Comment:
I put this question because I suppose that some polymerization can take place in the membrane through cluster formation, due to the low dielectric constant, which is a common phenomenon in coordination chemistry. I have never read about this, but PVC may be considered as a rather low dielectric surrounding. This could be another explanation phase.

Answer:
This is probably true for any other organic species, but it is a fact that protons are coming out of the membrane phase into the solution and there is definitely water in the membrane. In these solvent - polymeric membranes, there is about 0.1 M of water inside and even in heptane there is still about 0.001 M of water which is sufficient.

Question:
Some years ago there was a discussion about the possibility of fixed sites in these membranes. There was a paper by Keddam, and there was also a proposition that perhaps the solvent mediator was oxidized to form anions, or there were anions from impurities. Do you consider that the effect of these is a solved problem or worth considering any more?

Answer:
No, it is not a solved problem. We also tried to study this phenomenon a little further, and started to purify PVC very carefully and do not know yet what is going to happen with

those membranes. It seems probable that, in addition to what I said there are further anionic sites in the membrane and definitely the Keddam explanation was a very exciting one and it has not yet been disproved.

Question:
It has been mentioned that the selectivity order of the anions follows the Hoffmeister rule, and the order depends mainly on the hydration energy. It has also been said that in order to increase the selectivity one has to introduce chemical specificity. In this context some tin organic compounds were mentioned and their NMR spectra were shown. Could you make some comments on the nature of the compound formed which might be responsible for the increase in selectivity ?

Answer:
We had the most success with trialkyl tin chlorides. The problem with these is that they are slightly hydrolyzed, but recently we have prepared membranes which function properly. The NMR spectra show that when adding ions to a solution containing also these tin organic compounds something happens, namely that the coordination number changes to five from four, and even clusters may also be formed. By adding either bicarbon- ate or chloride ions, a shift occurs in the spectrum. This is an additional proof that a specific interaction occurs with the tin organic compounds. This has also been proved by vapour pres- sure osmometry.
The trick simply was to take a volatile solvent and to add the tin organic compound and an appropriate anion in order to bring it into solution. According to the suggestions of Dr.Pretsch we use kriptates which complex the cations. Using this trick we are able to show that certain tin organic compounds, e.g. octyl tin dichloride, interact both with bi- carbonate and chloride. This means that the number of particles decreases when the solutions are poured together.

Comment:
By using Mössbauer spectroscopy it would be really easy to prove whether any change in coordination number occurs and we would do the measurement for you with pleasure.

Answer:
The suggestion is very welcome, and we will send the compound immediately to Dr.Burger.

Question:
A correlation was given in the lecture between the partition coefficient of the ligand and the lifetime of the electrode. Certainly, the latter depends also on the conditions under which the electrode is employed. However, according to your opinion there exists a well defined correlation between these two parameters only.

Furtheron, could you explain why is the limit 5.5 ?

Answer:
As already said, certain assumptions were made concerning the lifetime of an electrode. However, the conclusion seems to be analytically relevant. If we take for example an Orion electrode from a new series, the correlation holds for that, or if we take a Philips electrode, it roughly holds for that, too. Of course, the area and volume of the membrane as well as the curvature of the membrane surface also play an important role in this respect. Supposing that the sample solution gets saturated with the carrier and after every 30 minutes the sample is replaced, and we work 24 hours a day, then we end up with a lifetime of one year if the lipophilicity of the carrier is about $10^{5.5}$.

Considering the compounds people work with as ionophores, one can expect a very short lifetime.

KEYNOTE LECTURES

X-RAY PHOTOELECTRON SPECTROSCOPY – APPLIED
TO INVESTIGATIONS OF COPPER(II) ION-SELECTIVE ELECTRODES

M. F. EBEL

Institute for Technical Physics, Technical University, Vienna, Austria

ABSTRACT

X-ray photoelectron spectroscopy is a surface technique applicable
to an information depth of 10 nm maximum. The informations supplied by
this method, theoretical considerations and appropriate evaluation procedures are well suited to obtain informations about the surface of
Cu(II) ion-selective electrodes.

INTRODUCTION

X-ray photoelectron spectroscopy (XPS) is basing upon the photoel-
tric effect: when material is exposed to electromagnetic radiation of sufficient high photon energy $h\nu$ the emission of electrons is observed. The
energetics of the process are defined by the Einstein relation /1/

$$E_k = h\nu - E_b$$

E_b is the binding energy of a certain species of electron in the material
and E_k is the kinetic energy with which such an electron is ejected. Figure 1 gives a scetch of the experimental situation. Due to the small
mean free path of the photoelectrons, which means small escape depth, XPS
is a s u r f a c e technique, applicable to an information depth of
10 nm maximum, supplying several informations about the surface of solids.

QUALITATIVE ANALYSIS

From a sample consisting of different elements different species of
electrons are ejected having various kinetic energies. The experimental de-

89

termination of the kinetic energy of the photoelectrons by an energy analyzer leads to a photoelectron spectrum. Figure 2 shows a photoelectron
spectrometer with spherical analyzer and without retarding field (GCA
McPherson ESCA-36). By means of a variation of the sphere voltage only
electrons of a certain energy can pass to the exit slit and thus enter the
detector. In Figure 3 the qualitative XPS-results of a lunar rock sample
are given /2/. Together with tabulated E_b-values /3/ the photoelectron
peaks corresponding to certain kinetic energies can be related to certain
species of electrons and give an identification of the elements. Furthermore, the energy levels occupied by electrons are quantized, consisting of
a series of discrete bands. That essentially reflects the shell character
of the electron structure of the sample.

Figure 4 illustrates the "elemental sensitivity" of the different
orbitals for Mg Kα excitation and for an instrument without retardation
/4/. Calculated countrates are depicted versus the atomic number. For some
elements (Z=11-26,29,38, in the neighbourhood of Z=55) the "elemental sensitivity" is poor, best "elemental sensitivity" is indicated for Z=79(Au).

QUANTITATIVE ANALYSIS

Applying XPS to quantitative analysis theoretical considerations have to be performed, including the influence of surface roughness and contamination. Recently, a theoretically based concept of fundamental parameters has been developed for quantitative XPS analysis without reference
samples /5,6/ and could be confirmed by experiments/5/. In Figure 5 the
mathematical expression is given, including the fundamental parameters
which have to be known for evaluation of measurements. Figure 6 depicts
the results of XPS-analysis without reference samples performed on a great
number of binary Ag-Au-, Ag-Cu-, Au-Cu- and ternary Ag-Au-Cu- alloys versus
the chemical analysis of the bulk material.

CHEMICAL SHIFT

The fact that the core electron binding energies of an atom normally
vary measurably with the change of the chemical environment perhaps has

aroused most interest in photoelectron spectroscopy among chemists. The effect was first observed in a routine study of sodium thiosulphate, showing a doubling of the sulphure core signals /3/. In contrast the XP-spectrum of sodium sulphate has a single sulphure 2p signal (Figure 7).

Figure 8 /7/ shows XPS studies carried out on a Cu(II) ion selective electrode (Mg Kα radiation, 5 kV, 40 mA). The following signals were measured: Cu 3p, Cu 2p$_{3/2}$, S 2p, O 1s, C 1s. Signals corresponding to the two oxidation states of sulphure could be detected, the one corresponding to sulphide is of higher intensity and appears at a higher kinetic energy, the other due to sulphate appears at a lower kinetic energy and has a smaller intensity.

In previous investigations the redoxsensitivity of Cu(II) ion-selective electrodes has already been detected /8/. 0,1 M KMnO$_4$ solution was used as oxidant and 0,1 M ascorbic solution as reductant. Oxidizing treatment was found to cause changes in the surface layer of the electrode membrane, which could be followed by checking the electrode function. As the oxidation goes forth the electrode performance gradually deteriorates as indicated by the reduction in the slope of the electrode calibration curve. After oxidation for a period of half an hour concentration changes of more orders of magnitude are hardly detected by the electrode, as shown in Figure 9 /7/. Regarding the XP-spectra of CuS electrodes after soaking in potassium permanganate the question arises according to which scheme CuS and CuSO$_4$ are grouped on the surface. Figure 10 shows the assumed possible surface structures.

In order to find a distinction between these different surface conditions argon-ion etching was performed. Figure 11 demonstrates how surface structures are altered by this treatment. The photoelectron spectra /9/ show that the sulphate signal decreases obviously whereas the sulphide signal increases. This result corresponds to the thickness distribution which meets the "reality".

VARIABLE TAKE-OFF ANGLE

A rotation of the sample around an axis parallel to its surface causes a variation of the take-off angle. This development of the XPS-method and an appropriate evaluation procedure allows the determination of the layer thickness of a thin surface layer /10,11/. Together with theoretical considerations it is possible to distinguish between elements of the surface layer and the substrate beneath. For the specific problem of CuS membrane electrodes a distinction between S 2p photoelectrons ejected from the sulphate layer and from the sulphide substrate is possible. For determination of the mean thickness of the sulphate layer the sulphide signal of the substrate was measured at 20 different take-off angles (see Figure 12). A least square fit /11/ leads to the reduced thickness D/λ-value of 0,26 (D=thickness of the sulphate layer plus a very thin contamination, λ=mean free path of the photoelectrons). Together with the assumption $\lambda = 30 \overset{o}{A}$ the total thickness of about 8 $\overset{o}{A}$ is determined.

SCANNING XPS

To gain more informations about the surface morphology a stepwise shifting of the sample within the spectrometer is necessary. These displacements direct different small areas of the sample towards investigation by the spectrometer and the ratio between sulphate and sulphide line intensities changes. Conclusions can be drawn concerning the frequency of the thickness distribution and furthermore, whether at the beginning of the oxidizing treatment areas of sulphate can be detected. The theory and the evaluation of the surface specific imaging technique have been described elsewhere /12/. First experiments employing the scanning device have been performed on a sample depicted in Figure 13. This figure also shows the results of the investigations. First measurements on CuS pellets gave "island diameters" of approximately 1 mm.

ACKNOWLEDGMENT

The photoelectron spectrometer has been purchased by financial support of the "Fonds zur Förderung der wissenschaftlichen Forschung in Österreich" (Projekt Nr. 1567).

REFERENCES

1. A.Einstein, Ann. Phys <u>17</u>, 132 (1905)
2. GCA McPherson, Acton, Massachusetts, customers communication
3. ESCA, Atomic, Molecular and Solid State Structure Studied by means of Electron Spectroscopy, by K.Siegbahn et al., Uppsala 1967
4. M.F.Ebel, Mikrochim. Acta Suppl. <u>8</u>, 115 (1978)
5. Surface and Interface Analysis, Vol.1, No.2, 58 (1979)
6. K.Hirokawa and M.Oku, Talanta <u>26</u>, 855 (1979)
   K.Hirokawa, T.Sato and M.Oku, Z. Anal. Chem. <u>297</u>, 393 (1979)
7. M.F.Ebel, Surface and Interface Analysis, in press
8. J.Pick, K.Tóth and E.Pungor, Anal. Chim. Acta <u>61</u>, 169 (1972)
9. E.Pungor, K.Tóth, M.Pápay, L.Polos, H.Malissa, M.Grasserbauer, E.Hoke, M.F.Ebel and K.Persy, Anal. Chim. Acta <u>109</u>, 279 (1979)
10. R.J.Baird and C.S.Fadley, J.Electron Spectrosc. Relat. Phenom. <u>11</u>, 39 (1979)
11. M.F.Ebel, J.Electron Spectrosc. Relat. Phenom. <u>14</u>, 287 (1978)
12. M.F.Ebel, W.Gröger and E.Pungor, Hungarian Sci. Instr. (Special Issue for the occasion of the Third Scientific Session on Ion-Selective Electrodes, Mátrafüred, 1980)

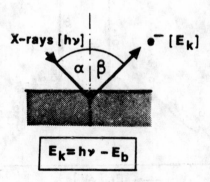

Figure 1. Emission of photoelectrons by means of X-ray excitation and description of the process by the Einstein relation /1/

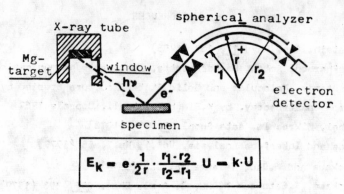

$$E_k = e \cdot \frac{1}{2r} \cdot \frac{r_1 \cdot r_2}{r_2 - r_1} \, U = k \cdot U$$

Figure 2. X-ray photoelectron spectrometer (GCA McPherson ESCA-36).
The kinetic energy of the photoelectrons is proportional to
the sphere voltage

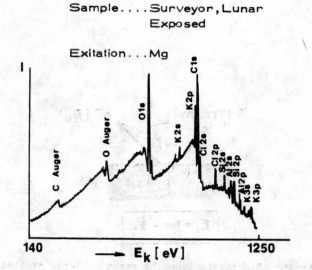

Figure 3. Photoelectron spectrum excited by Mg Kα radiation for qualitative analysis of the elements of a lunar rock sample /2/

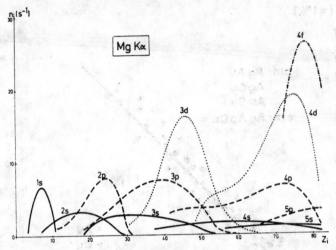

Figure 4. Calculated photoelectron countrates ($n\ s_{1/2}$, $n\ p_{3/2}$, $n\ d_{5/2}$, $n\ f_{7/2}$) in arbitrary units, for Mg Kα radiation and for an instrument without retardation

$$n_i = K \cdot E_{i\ kin}^k \cdot \frac{\sigma_i^k}{A_i} \cdot (1 + \frac{\beta_i^k}{4})\ \frac{c_i}{\sum \frac{c_j}{\rho_j \cdot \Lambda_{ij}^k}}$$

n_i ..... photoelectroncountrate

K ...... constant

σ_i^k ..... photoabsorption cross–section (Scofield)

A_i .... atomic weight

β_i^k .... asymmetry parameter (Reilman et al)

c_i ..... concentration in wt %

ρ_j ..... density

Λ_{ij}^k ... electron mean free path (Penn)

Figure 5. Mathematical expression for the quantitative evaluation of XPS results, employing fundamental parameters instead of reference samples /5/

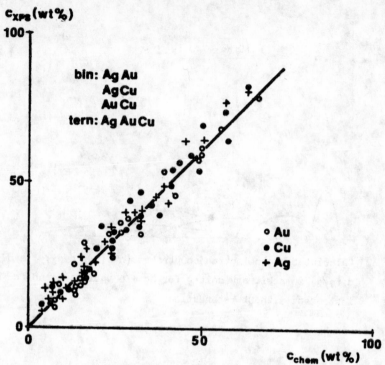

Figure 6. XPS concentration of Ag, Au, Cu obtained by quantitative XPS analysis without reference samples versus the results of the chemical analysis for ternary and binary alloys of the Ag—Au—Cu—system /5/

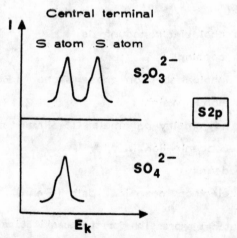

Figure 7. Variation of core electron binding energies with change of chemical environment (chemical shift) investigated for sodium—thiosulphate /3/ and sodiumsulphate

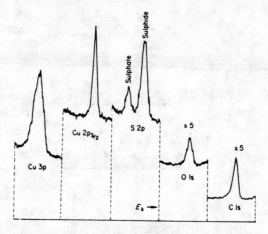

Figure 8. XP-spectrum of an untreated CuS electrode /7/

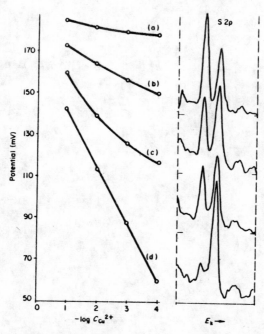

Figure 9. Cu(II) response and corresponding XP-spectrum of a CuS electrode after soaking in $KMnO_4$ for (b) 10 min and (a) 30 min. The electrode treated for 10 min with $KMnO_4$ was regenerated by ascorbic acid for (c) 10 min and (d) 30 min /7/

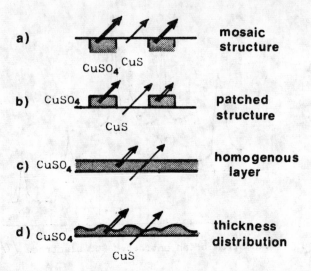

a)   mosaic structure

$CuSO_4$ CuS

b) $CuSO_4$ patched structure

CuS

c) $CuSO_4$ homogenous layer

d) $CuSO_4$ thickness distribution

CuS

Figure 10.   Assumed possibilities for the occurence of sulphate on the surface of CuS electrodes

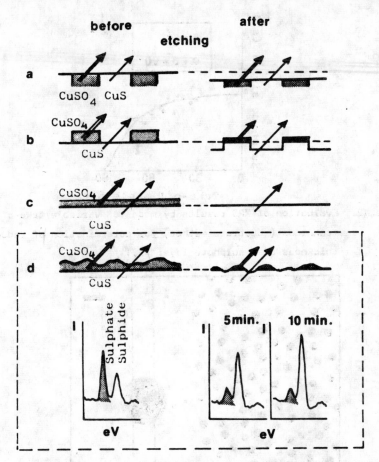

Figure 11. Alteration of the surface structure by argon-ion treatment
(a,b,c,d as in Figure. 10). The XPS results /9/ correspond
to thickness distribution

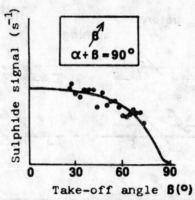

Figure 12. Evaluation of XPS results by means of variable take-off angle technique (substrate method) for determination of reduced thickness of a sulphate layer /9,11/

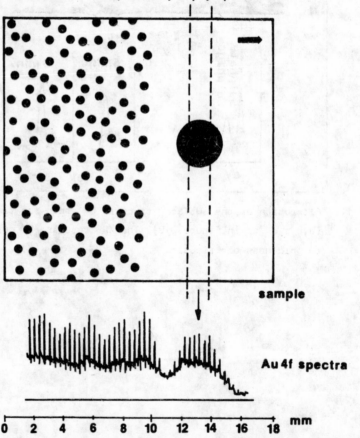

Figure 13. Basic experimental arrangement and corresponding measured XP-spectra for the determination of Au-island diameters by means of a stepwise shift of the sample. The information is gained from the meanvalue and the standard deviation of the Au 4f peak heights /12/

QUESTIONS AND COMMENTS

Participants of the discussion: J.D.R.Thomas, E.Pungor,
J.Siemroth, E.Lindner, K.Burger, A.Lewenstam, M.Ebel

Question:
We have been engaged with the study of copper ion-selective
electrode membranes, but we did not get the results we had
hoped. Did you look at copper ion-selective electrode membranes
exposed to potassium chloride in copper/II/ nitrate, or any
other kind of combination of exposure to chloride, with which
you have peculiar surface effects ?

Answer:
All the electrodes were prepared and pretreated by Dr.Pungor's
group, but none of them was exposed to chloride interferant.

Comment:
We have studied the same membrane system you did, by X-ray
diffraction technique, and the results will soon appear in the
Journal of Electrochemical Society. After pretreatment we have
found sulphur and chloride at the surface of a pure copper/II/
sulphide membrane.

Answer:
The same has been found by us, and was proved by chemical
methods, too. The bulk of the membrane material was copper/II/
sulphide, but by these methods copper/I/ could not be observed,
although it is most probably present.

Comment:
Some work was done by Japanese workers on copper minerals by
photoelectron spectroscopy. In several cases they found exclusively copper/I/ sulphide to be present, e.g. in chovalite.

Question:
We are interested in finding out what happens at the surface
of electrodes, e.g. at an iodide-slective electrode. My
question is whether you can distinguish between the different

structures of precipitates formed on the surface of silver
iodide, e.g. silver bromide or the mixed crystals of silver
iodide and bromide as described by Jaenicke in the fifties.
We have found that the silver iodide based electrode covered
with electrodeposited silver bromide shows response characteristics entirely different from those of the electrodes soaked
in bromide solutions.

Answer:
I have already done some experiments in this field, and I am
going to do the evaluation of the data very soon.

Comment:
In addition to the previous comment I mention that the bromide
signal observed for the iodide electrode covered with silver
bromide only came from the surface. It appears that the nature
of the signal depends on the type of pretreatment.

Question:
Can you distinguish between potassium bromide adsorption and
silver bromide formation at the electrode surface ?

Answer:
I hope I will be able, as it is a question of the variation of
the take-off angle.

Comment:
It is very important to distingnish between the adsorption
and dissolution precipitation mechanism at the formation of
a new phase. There is a mixibility of bromide and iodide or
bromide and chloride within the phase, and it is extremely
important whether the new phase partly or completely covers
the active surface phase of the electrode.

SELECTIVITY PROBLEMS OF ION-SELECTIVE ELECTRODES

A. HULANICKI

Department of Chemistry, University of Warsaw, Warsaw, Poland

ABSTRACT

Classification of various types of interferences in measurements with ion-selective electrodes has been discussed.
This includes four types of effects: selectivity interferences
/ion-exchange competition/, specific chemical interferences,
non-specific interferences and solution interferences. The
generalized model for selectivity interferences in solid-
state electrodes has been presented and discussed.

INTRODUCTION

The starting point in developement of theory and practice
of ion-selective electrodes seems to be the formulation of
the so called Nikolsky equation [1], which described the behaviour of potentiometric sensors in solutions reflecting
real analytical conditions, i.e. when beside the main ion
other species are present which interfere in the analytical
potential measurements. The semiempirical Nikolsky equation
is valid for various types of ion-selective electrodes independently on the detailed mechanism of electrode action, and
therefore can be applied to glass, precipitate, liquid ion-
exchange or neutral carrier electrodes.However, it should be
always remembered, that Nikolsky equation concerns only one
type of interferences in analytical determinations. In prac-

8*

tical applications several species present in the analyte solution can exert a harmful effect on final results and therefore it seems useful to state clearly various types of processes which may disturb the analytical application of ion-
selective electrodes. Such effects can be divided into four
main types. These are:
- interferences of ions of the same charge type which interact with the electrode according to the same principal mechanism as the main ion. Such effects are ion-exchange
  competition effects or may be shortly named as selectivity
  interferences.
- interferences of various species /ions, molecules/ which
  enter into different types of chemical processes with the
  membrane material. These processes were observed for several substances, but they differ in degree and mostly such
  effects are specific for a given electrode. They may be
  described as specific chemical interferences.
- interferences in measurements often occur due to chemical
  or physical phenomena at the electrode solution interface,
  or due to changes in physical properties of the sample solution. Such effects are mostly non-specific interferences.
- the last type of interferences are usually recognized as
  solution interferences and are caused by changes of the
  chemical form of the ions sensed by the electrode.

CHARACTERISTICS OF VARIOUS TYPES OF INTERFERENCES

The selectivity interferences, because of the require-
ment of similarity of the mechanism of ion interaction with
the electrode are exclusively cationic for cation-selective
electrodes and vice versa anionic for anion-selective electrodes. Therefore presentation of selectivity coefficients
for anions in the case of cation-selective electrode seems
not be proper. Most generally the ion-exchange reaction is
represented by the equation

$$SI + J \rightleftharpoons SJ + I \tag{1}$$

where S represents charged or neutral active sites in the membrane, I - the main ion, J - the interfering ion. The qualitative character of the selectivity depends on the nature of exchange sites in the membrane which varies with the type of the material used. For glass electrodes the sites are ionogenic silicate groups, modified by the added components of glass which change the position of the equilibrium [2,3] according to the ion-exchange constant.

Concerning the precipitate based electrodes in the ion-exchange reaction participate two insoluble substances which interact with the ions in solution. The necessary condition for electrode activity and prediction of interference of the ion-exchange competition type is the insolubility of SI and SJ which can be guessed from the values of solubility products of respective substances [4].

Following this path the exchange reaction constant is given by the solubility products ratio, but as it will be shown, only in special cases this directly describes the extent of selectivity [5].

In liquid-state membrane various types of processes are responsible for selectivity interferences depending on the properties of the active material in the membrane. On one hand   ion-exchangers with positive or negative charge form as a consequence a neutral complex or ion-pair in the membrane phase. On the other hand a neutral carrier gives rise to formation of charged species in the lipophylic phase. Depending on the complex properties, which are modified by the solvent, /plasticizer, mediator/ the species in the organic phase may be to a different degree dissociated or associated. Without entering into details it can be stated that the position of the equilibrium of reaction (1) depends primarily on the ratio of extraction coefficients of two ions in question [6-12]. For various types of membranes this can be expressed by some more exactly defined physicochemical parameters, including complex stability constants in one or two phases, ion partition coefficients, solvation energies etc. Those parameters depend on the composition and properties

105

of the liquid membrane phase as specific solvent-solute interactions, dielectric constant and others. Therefore several papers were devoted to the study of the effect of those
properties on the variation of selectivity. As was indicated
in the case of solid-state electrodes the equilibrium parameters are often unsufficient for quantitative description
of selectivity but can give useful qualitative informations
about the existence of interferences. For a complete picture
a knowledge of transport phenomena and parameters describing
it is often required [13-17]. It should be however indicated
that in some cases these parameters are cancelled, since
they appear in equations as ratios for two physically similar
species and consequently the values of such ratios are close
to unity. When this occurs the membrane potentials were found
to be additive which in turn results in additivity of selectivity coefficients [18,19].

Many others report significant discrepancies in selectivity coefficient values for liquid-state electrodes. This
often follows from facts which are difficult for precise description or may even disappear from the sight of a not sufficiently careful observer. A significant source of errors
may result due to gradual changes of the liquid membrane composition[20]. This is connected with irreversible reactions
of the active membrane materials or with a selective strip-
ping of membrane components during prolonged contact with
aqueous solution or selective evaporation when keeping the
electrode dry.

Apparently false results may also arise when the total
concentrations of the main and interfering ions are drastically low or high in one of the two phases[19]. In these cases the initial conditions at the interface are seriously
modified by passage of ions from or to the other phase. As
a result the measured potential values and calculated on
their basis selectivity coefficients do not correlate with
the thermodynamic characteristics of the system and bulk
concentrations in the solution. This is reflected by the
non-parallel course of calibration curves in the low total

concentration range[18,21]. (Fig. 1).

The specific chemical interferences occur when the interfering substance with or without presence of the main ion is interacting with the membrane material modifying the electrode surface. The most widely discussed case of this type is the chloride interference on the copper selective electrode. A number of papers[22-24] has been devoted to elucidate the nature of such interference and most probably it is connected with a coupled redox and complexation process according to the following equations:

$$Ag_2S + 2Cu^{2+} + 6Cl^- \rightleftharpoons 2AgCl + 2CuCl_2^- + S^o$$
$$CuS + Cu^{2+} + 4Cl^- \rightleftharpoons 2CuCl_2^- + S^o$$
$$\text{or} \quad Cu_2S + 2Cu^{2+} + 8Cl^- \rightleftharpoons 4CuCl_2^- + S^o$$

These reactions result in passivation of the electrode or its transformation /in the case of Ag_2S, CuS membrane/ into a chloride sensitive electrode.

Another example may be the effect of oxidants present in solution, as iron(III) or molecular oxygen which oxidize the sulphide material of the electrodes[25]

$$MS + 2Fe^{3+} \rightleftharpoons M^{2+} + 2Fe^{2+} + S$$

or even lead   to sulphate formation. Such effects should be also considered as non-specific interference.

In the case of liquid-state electrodes specific chemical interferences occur when cation selective electrodes are disturbed by the presence of some anions, especially those which have low hydration enthalpy and therefore exhibit a more pronounced tendency to penetrate into organic phase. In neutral carrier electrodes such interference occurs because   electroneutrality of both phases should be preserved and it can be significantly eliminated using less polar organic phase or by incorporation of a lipophylic anion as tetraphenylborate in the membrane composition[26,27]. Similar effects were also found for calcium selective electrodes based on dialkylphosphate exchangers [28,29]. As media-

tors in such electrodes are used dialkylphenylphosphonates
which are not inert towards calcium ions and coordinate with
them forming in the membrane an oxonium ion serving as an
anion exchanger

$$\begin{matrix} R\!-\!O \\ R\!-\!O \end{matrix}\!\!\!> \underset{\underset{C_6H_5}{|}}{P}\!=\!O \; + \; Ca^{2+} \longrightarrow \left[\begin{matrix} R\!-\!O \\ R\!-\!O \end{matrix}\!\!\!> \underset{\underset{C_6H_5}{|}}{P}\!=\!O \longrightarrow Ca \right]^{2+}$$

Such species present in the membrane show anion response in
presence of perchlorate, thiocyanate and some other more lipophylic anions (fig. 2).

Other examples of specific interferences are observed
when solution of analyte contains substances which complex
one of the ions forming soluble product. In such processes
the electrode surface is corroded and successively destroyed,
but initially such reactions may be used as a basis of indirect potentiometric procedures of determination. As example
may serve the afore mentioned effect of iron/III/ as oxidant
on the copper sulphide electrode which is the basis of iron
/III/ determination[25]. Chemically different type of reaction occurs in presence of cyanide with the silver iodide
electrode and finds application in cyanide determination[30,
31,32]. Also mercury/II/ ions were determined through their
interaction with silver iodide electrode[32]. A number of
processes belong to this category including those when the
membrane components are complexed by the ligands present in
solution /e.g. AgCl - NH_3 interaction/.

The effect of surfactants seems to be a very complex one.
They can physically poison the membrane as happens in the
case of dimethyldidecylammonium bromide influencing the potassium selective valinomycin PVC electrode[34,35]. Such behaviour should be considered as a non-specific interference,
however very often the effect of surfactants reflects the
influence of kinetic parameters on electrode response. In
the case of another ammonium salt-type surfactant, Hyamine,
the interference in calcium determination with the liquid
ion-exchange electrode is a typical selectivity interferen-

ce, in which the surfactant cation competes with calcium for the sites in the membrane [36].

Another mechanism was supposed when the effect of a linear alkylbenzene sulphonate on calcium electrode was investigated [36]. The assumed mechanism includes monolayer formation and following extraction which introduces new sites into membrane. Addition of surfactants forms a mixed-site exchanger which in turn is immune to such interferences. Such processes should be considered as non-specific interferences.

The non-specific interferences include a variety of different phenomena which are probably not exactly known and little can be said generally about them. Several such effects are eliminated by very careful matching of measurement conditions for the analyte and for standards. Serious difficulties are connected with adsorption phenomena at the electrode active surface in presence of proteins or other organic constituents. The presence of colloidal particles and particulate matter which influences not only the membrane potential at the electrode surface but also effect the potential reading through the change of liquid-juncton potential [38]. To very troublesome processes belong chemical passivation effects, which were at least partly mentioned among the specific chemical interferences. The known example presents the lead selective electrode which depending on its composition and preparation loses its sensitivity towards changes of lead ion concentration. Such processes are connected with the sulphide oxidation [25,39] even by traces of oxidants. Attempts to eliminate their interference by a differential system with two similar electrodes failed, but some improvement was achieved in flow determination of sulphate when both differentially connected electrodes were cathodically polarized with the 15 nA current during the measurements [40]. It seems that such and similar effects are at least partially eliminated when differential measurement with ion selective electrodes were more widely applied.

It has been also found that the behaviour of several metal-selective solid-state electrodes is disturbed by the pre-

sence of polyaminopolycarboxylate ions as EDTA or NTA. In the
presence of excess of such ligands the lead, cadmium and especially copper electrodes do not respond as follows from
solution equilibria calculation but the potential is significantly more positive[41,42]. (Fig. 3). It was assumed that
such behaviour is connected with an adsorption mechanism [43]
but until now no definite evidence has been presented. It
must be pointed out that similar interference is not observed in the case of polyamine ligands as TRIEN or TETREN. The
main difficulty with those phenomena is that the extent of
interferences is only qualitatively predictable and until now
any quantitative evaluation of interference is possible.

The last type of interferences are in fact pseudo-inter-
ferences because the electrode detects properly   only one
species in solution. The apparent interference is observed
in the case when the total concentration of analyte remains
constant but change the mole fractions of different forms.
The processes which are responsible for such pseudo-interfe-
rences are mainly protolytic or complexation reactions and
occurence of such interferences is the basis of application
of ion-selective electrodes for investigation of ion equilibria in solution. On the other hand known equilibrium constants and relevant concentrations enable quantitative estimation of degree of interference. Because also among those
processes the kinetic parameters for heterogenic reactions
play an important role, equilibrium   data may give in especially unfavourable cases only a rough estimation of interferences.

SELECTIVITY COEFFICIENTS OF SOLID-STATE ELECTRODES

A significant scatter of selectivity coefficient data for
several apparently well-known and simple systems, as for the
silver chloride membrane with interference of bromide, has
stimulated us to a more detailed study of the mechanism of
selectivity.

The electrode behaviour according to the total equili-

brium model as suggested by Pungor and Toth [3] assumes that
the selectivity coefficient should be equal to the ratio of
solubility products of respective salts, i.e. in the case of
bromide interference on the chloride electrode:

$$K_{Cl,Br}^{pot} = \frac{K_{so\ AgCl}}{K_{so\ AgBr}} = 330$$

This value has been found experimentally by many authors and
it corresponds to the final equilibrium between two insoluble
salts and their ions in solution. It was assumed that the two
salts form a mixed phase being electrode active, and usually
such state exists when the ion concentrations are high and
their interaction is sufficiently long.

In the diffusion layer model developed in our laboratory
[5,44] for a system of two insoluble salts participating in
a metathetic reaction it was assumed that the surface concentrations are not equal to those in the bulk solution. Addi-
tionally it should be supposed that the newly formed phase
is not electrode active. For such system, using the diffusion
equations for the ion-flow, the selectivity coefficient is
given as the ratio of diffusion coefficients

$$K_{Cl,Br}^{pot} = \frac{D_{Br}}{D_{Cl}}$$

This expression is however valid only in the case when the
exchange constant for the reaction considered is greater than
unity. In the opposite case /i.e. $K < 1$/ selectivity coefficient equals as previously to the ratio of solubility products. This gives the answer why for chloride interferences
in bromide selective electrode the selectivity coefficients
given by various authors have always nearly the same value.

It can be shown that these two models are two limiting
cases of a generalized model. When the thermodynamic force
is small, or when it is acting for a limited period of time,
then the transport phenomena are predominant. In opposite
conditions when the concentrations are large and the time of

111

interaction is sufficiently long the equilibrium is attained and the selectivity coefficients are expressed through solubility products. There is possible to present an exact description of the variability of selectivity coefficients when the apparent coverage factor $s = No(AgBr)$ has been introduced, which corresponds to mole fraction of the active surface sites composed of the new phase - in the discussed case - of silver bromide.

The equation has the following form:

$$E = \text{const} - \frac{RT}{F} \ln \frac{1}{s \dfrac{D_{Br}}{D_{Cl} K_{Cl,Br}} + (1-s)} \left([Cl^-] + \frac{D_{Br}}{D_{Cl}} [Br^-] \right)$$

and it can be shown that it includes the former equations as limiting cases.

When the potential of an ion-selective electrode is observed as a function of time in various experimental conditions records were obtained which support the presented relationships (Fig. 4). They show that the potential change is more rapid when the ion concentrations in solutions are greater, when the temperature is higher or when the solution in contact with electrode surface is vigorously stirred.

The comparison of the apparent coverage factor in the presented equation with the experimentally measured time of the electrode response as function of the electrode potentials shows that the coverage factor is proportional to the square root of time. This corresponds to phenomena which are observed for example in chronopotentiometric conditions.

The behaviour of the bromide-selective electrode in chloride solution is also in good agreement with the theory and the electrode potential does not vary with time as a function of experimental conditions. It follows that for such electrode most selectivity coefficients are consistent (Fig.5).

The case of silver chloride or silver bromide electrodes the processes occuring in solution and at the electrode surface. There is no need to consider ion transport within the membrane which is negligible, and its contribution to the total electrode potential is unsignificant. The situation is different when migration of ions takes place in the membrane phase. Such case exists for the fluoride-selective electrode and hydroxyl ion interference. The migration of fluoride ions to the electrode surface and of hydroxyl ions into the electrode give rise to the diffusion potential and must be taken into account in the general consideration of electrode behaviour. By formal analogy a site-filling factor can be introduced in those consideration, which corresponds to the mole fraction of sites exchanged for the interferant ions. This gives a somewhat more complicated equation:

$$E = const - \frac{RT}{F} \ln \frac{(1-s) + s \frac{U_{OH}}{U_F}}{\frac{s}{K_{F,OH}} \frac{D_{OH}}{D_F} + (1-s)} \left([F^-] + \frac{D_{OH}}{D_F} [OH^-] \right)$$

which qualitatively has been experimentally verified. However exact comparison of calculations with theory is in this system not possible quantitatively because the accurate value of the exchange constant is not known, and even we have no arguments for supporting the wiev that for one lanthanum atom one or more fluoride ions are exchangable. This follows from large discrepancies of the solubility products of lanthanum fluoride and hydroxide, but using roughly estimated values relatively good agreement is obtained. Also in this case experimental conditions which influence of diffusion rate were tested and the expectations were fulfilled (Fig.6).

The study of selectivity and interferences of ion-selective electrodes enables not only a deeper insight into the mechanism of electrode action but also makes possible a better utilisation of electrodes in practical applications.

REFERENCES

1. B.P.Nikolsky, Acta Physicochim., URSS, $\underline{7}$, 597 /1937/
2. G.Eisenman, Glass Electrodes for Hydrogen and Other Cations, Principles and Practice, M.Dekker, New York, 1967
3. B.P.Nikolsky, M.M.Schulz, A.A.Beljustin, Wiss. Z. TH - Leuna-Merseburg $\underline{18}$, 573 /1976/
4. E.Pungor, K.Tóth, Anal.Chim.Acta, $\underline{47}$, 291 /1970/
5. A.Hulanicki, A.Lewenstam, Conf. on Ion-selective Electrodes, Budapest, 1977, p.395
6. S.Bäck, Anal.Chem., $\underline{44}$, 1696 /1972/
7. S.Bäck, J.Sandblom, Anal.Chem., $\underline{45}$, 1680 /1973/
8. A.Jyo, Mihara N.Ishibashi, Denki Kagaku, $\underline{44}$, 268 /1976/
9. R.P.Scholer, W.Simon, Helv.Chim.Acta, $\underline{55}$, 1801 /1972/
10. R.P.Scholer, W.Simon, Chimia $\underline{24}$, 372 /1970/
11. G.Baum, J.Phys.Chem., $\underline{76}$, 1982 /1972/
12. H.J.James, G.P.Carmack, H.Freiser, Anal.Chem., $\underline{44}$, 853 /1972/
13. J.P.Sandblom, G.Eisenman, J.L.Walker jr., J.Phys.Chem., $\underline{71}$, 3862 /1967/
14. K.Selinger, Thesis, Academy of Medicine, Wrocław, 1979
15. C.Fabiani, Anal.Chem., $\underline{48}$, 865 /1970/
16. C.Fabiani, P.R.Danesi, G.Scibone, B.Scuppa, J.Phys.Chem., $\underline{78}$, 2370 /1974/
17. W.E.Morf, W.Simon, 2nd Symp. on Ion-selective Electrodes Mátrafüred, Akadémiai Kiadó, Budapest, 1977
18. A.Jyo, M.Torikai, N.Ishibashi, Bull.Chem.Soc. Japan $\underline{47}$, 2862 /1974/
19. N.Yoshida, N.Ishibashi, Chem.Lett., 497 /1974/
20. A.Hulanicki, Z.Augustowska, Anal.Chim.Acta $\underline{78}$, 261/1975/
21. A.Hulanicki, R.Lewandowski, Chemia anal./Warsaw/ $\underline{19}$, 53 /1974/
22. P.Lanza, Anal.Chim.Acta $\underline{105}$, 53 /1979/
23. J.C.Westall, F.M.M.Morel, D.N.Hume, Anal.Chem., $\underline{51}$, 1792 /1979/
24. J.Gulens, Ion-Selective Electrode Revs. /1981/, in print
25. Y.S.Fung, K.W.Fung, Anal.Chem., $\underline{49}$, 497 /1977/

26. W.E.Morf, G.Kahr, W.Simon, Anal.Lett., 7, 9 /1974/
27. J.H.Boles, R.P.Buck, Anal.Chem., 45, 2057 /1973/
28. J.Růžička, E.H.Hansen, J.Chr.Tjell, Anal.Chim.Acta 67, 155 /1953/
29. A.Hulanicki, M.Trojanowicz, Z.Augustowska, Chemia anal. /Warsaw/ 26, /1981/, in print
30. K.Tóth, E.Pungor, Anal.Chim.Acta, 51, 221 /1970/
31. I.Sekerka, J.P.Lechner, Water Res., 10, 479 /1976/
32. G.P.Bound, B.Fleet, H. von Storp, D.H.Evans, Anal.Chem., 45, 788 /1973/
33. W.E.Morf, G.Kahr, W.Simon, Anal.Chem., 46, 1538 /1974/
34. S.M.Hammond, P.A.Lambert, J.Electroanal.Chem., 53, 155 /1974/
35. R.A.Durst, Clin.Chim.Acta, 80, 225 /1977/
36. R.A.Llenado, Anal.Chem., 47, 2243 /1975/
37. R.A.Durst, in Ion-Selective Microelectrodes, /Eds. H.J. Berman, W.C.Hebert/, Plenum Press, New York, 1974
38. H.Jenny, T.R.Nielsen, N.T.Coleman, D.E.Williams, Science, 112, 164 /1950/
39. G.Johansson, K.Edström, Talanta 19, 1623 /1972/
40. M.Trojanowicz, Anal.Chim.Acta, 114, 293 /1980/
41. R.Blum, H.M.Fog, J.Electroanal.Chem., 34, 485 /1972/
42. G.Nakagawa, H.Wada, T.Hayakawa, Bull.Chem.Soc.Japan 48, 424 /1975/
43. J.M.Heijne, W.E. van der Linden, Anal.Chim.Acta 96, 13 /1978/
44. A.Hulanicki, A.Lewenstam, Talanta 24, 171 /1977/.

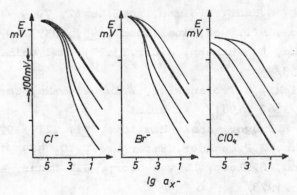

Fig. 1. Calibration curves for anion-selective electrodes /Aliquat-PVC/. The thick lines correspond to the response to the main anion incorporated in the membrane, Cl^-, Br^- and ClO_4^-, respectively. The thin lines represent calibration curves for the interfering ions in the order: Cl^-, Br^-, I^-, ClO_4^-.

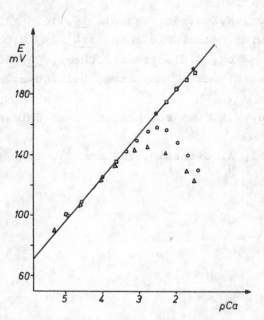

Fig. 2. Anionic interferences on calcium selective electrode:
● - Cl^-, □ - NO_3^-, ○ - SCN^-, △ - ClO_4^-.

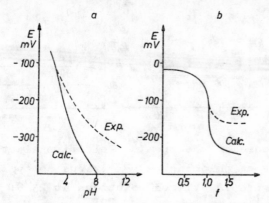

Fig. 3. Copper selective electrode response in presence of
excess of EDTA. a. — pH dependence for the 1:1 buffer
CuEDTA/EDTA, b. — titration curves for compleximetric
determination of copper. The dashed lines represent
calculated curves on the basis of CuEDTA stability
constants.

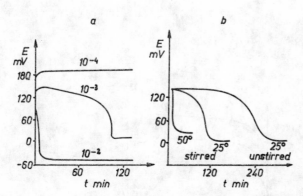

Fig. 4. Response of the chloride selective electrode in presence of bromide. a. — Various concentrations of bromide in stirred solutions at 25°C. The final potential for 10^{-4} mol/1 bromide corresponds to $K^{pot}_{Cl,Br}=2$,
for 10^{-3} and 10^{-2} mol/1 corresponds to
$K^{pot}_{Cl,Br}=330$. b. — Effect of stirring and temperature
in the case of 10^{-3} mol/1 bromide solutions.

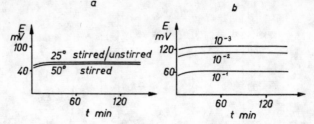

Fig. 5. Response of the bromide selective electrode in presence of chloride. a. - Various concentrations of
chloride in stirred solutions at 25°C. The final potentials correspond to $K^{pot}_{Br,Cl}$=0.003. b. - Effect of
stirring and temperature in the case of 10^{-1} mol/l
chloride solutions.

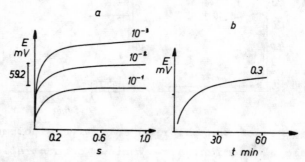

Fig. 6. a. Calculated response of the fluoride selective
electrode for indicated exchange constants as a function of site filling factor, s,
b. - Experimental curve corresponding to initial value of $K^{pot}_{F,OH}$=1, and final value $K^{pot}_{F,OH}$=0.3.

QUESTIONS AND COMMENTS

Participants of the discussion: J.Koryta, E.Pungor, R.P.Buck,
W.Simon, A.Hulanicki

Comment:
This comment concerns the presentation of selectivity data,
more exactly, the presentation of the ratios of the diffusion
coefficients. The approach used is the Nernst layer concept
for describing the transport process. However, it is not
so simple since in all cases solved e.g. by Levich in chemical
hydrodynamics, the layer thickness is always a function of
the diffusion coefficient. A typical case where we have a
constant Nernst layer thickness is the rotating disc electrode.
The flux is proportional to the 2/3 power of the diffusion
coefficient. However, we can not take simply the diffusion
coefficient but have to introduce some operational exponents,
say α , β etc., because the diffusion coefficients are not
very definite, particularly in organic phases.
The other problem is connected with the diffusion in the
organic phase if you use e.g. a plastic film electrode. We
have a polymer matrix with pores where some droplets of liquid
join to each other. In such capillary systems, in my opinion,
it is not justified to use the Nernst concept at all, because
we must take into consideration the so called quiescent
diffusion also.

Answer:
In our paper the diffusion in the organic phase has not been
discussed. The diffusion layer concept was used only in
connection with solid-state electrodes. Furtheron, a simple
model of linear diffusion to the membrane has been assumed.
By solving the Fick equations, a relationship was obtained
in which the diffusion coefficients are involved. Maybe it is
not quite exact, but the agreement with experimental data for
the chloride electrode in the presence of bromide was quite
good showing that this is a case where we are close to solve
the problem.

Comment:
This comment is connected with electrodes which have a selectivity coefficient higher than one. In my opinion, such
electrodes are not selective to the first species involved.
So let us leave out of the discussion of ion-selective
electrodes such cases where the selectivity coefficents
exceed one. For example, if a silver chloride based electrode
is dipped into a solution containing bromide ions, then, after
a certain period of time the electrode surface becomes covered
with silver bromide, which means that it is not a chloride-
-selective electrode any longer. In my opinion, not the
selectivity is to be investigated but the physical processes
like transport phenomena, as discussed by Dr.Koryta. Unfortunately, selectivity coefficient data exceeding one are reported
in a number of papers. However, these sensors can not be
considered as selective electrodes in the presence of such
species, and we really do not know which species are respon -
sible for the electrode function under such conditions.

Answer:
After a sufficiently long time the silver chloride surface
will be covered by silver bromide. At the beginning, however,
we have the diffusion of both chloride and bromide to and
from the electrode surface.

Question:
What is the rate of exchange ?

Answer:
We do not know the heterogeneous exchange rates but they are
involved in the curves showing the effect of parameters like
concentration, time etc.

Question:
How can you calculate the concentrations in the surface
layer without knowing the exchange rate ? The potentials
measured are mixed potentials.

Answer:
The selectivity coefficients are experimentally measured data.

Comment:
The data are perfectly all right. However, it should be underlined that these should not be called selectivity coefficients,
since, in my opinion, a selectivity coefficient is defined as
having a value lower than one.

Answer:
The real meaning of these data is non-selectivity coefficients.

Comment:
I am very sorry to say that I completely disagree with these,
because you can get into serious trouble with liquid membrane
electrodes.
We have for example the radium electrode and we do not know
whether any other ion would give an electrode function. We
feel, however, that barium would do so. And really, barium is
found to be discriminated by the electrode by a factor of ten
over radium.

We have the valinomycin based potassium electrode with a
selectivity coefficient more than one in the case of rubidium
or caesium interfering ions and we still call it a potassium
electrode. This really puts us into problems.

Answer:
My comment was probably misunderstood. I repeat: there is an
ion to which the electrode responds, and there is another ion
to which the electrode shows a selectivity coefficient less
than one. In this case, for this pair of ions, the electrode
is selective to the former ion. I stress again that there is
no use of speaking of selectivity when the selectivity coefficient is larger than one, especially in the case of solid
electrodes.

Comment:
There is a distinction in this whole field between reactions
that are irreversible or taking very long time to reverse and
those which can be reversed quite easily. The case discussed

in the paper is an example which is very difficult to reverse. Therefore I would agree that it is not a good idea to talk about selectivity coefficients greater than one in that case. If I use a chloride electrode in bromide solution and put it back in chloride, it would not be the same electrode even if it does convert to chloride it is not going back that way. But if I go from nitrate to perchlorate with the tetraalkylammonium salt, I can in a reasonable time, flush out all the nitrate with perchlorate, and just because the perchlorate seems to have a bigger selectivity number than nitrate, I would think that as an indication that perchlorates are preferred over nitrates. I do hope that my method resulting in a selectivity number greater than one is reliable. If it were not, I would get into trouble in making the perchlorate electrode measure the nitrate interference, for which case the selectivity coefficient is less than one, and in taking the reciprocal of the data and saying that it was a number greater than one. That would not apply if the process were not reversible.

Comment:
I would like to draw your attention back to the problem of diffusion, and comment the applicability of Fick's linear diffusion equations for longer times. Unless you have a capillary system, you must take into consideration the role of natural convection. For example, if we carry out a classical chronoamperometric experiment with iron/III/, by setting the potential to the limiting current region, then the current vs. time curve follows exactly the $t^{-1/2}$ function at the beginning, but if we take about 100 seconds, then the current approaches its steady-state value, corresponding to the natural convection. I can refer to Dr.Ebel's review published about 15 years ago. I used his method in the case of a cyanide electrode for working out a β parameter instead of taking the ratio of the diffusion coefficients of cyanide and iodide.

Answer:
I agree completely with the strict conditions for the applicability of the laws of diffusion, it is quite clear.

SOME RECENT IMPROVEMENTS IN ION-SELECTIVE ELECTRODES

J. D. R. THOMAS

Chemistry Department, Redwood Building, UWIST, Cardiff, Wales

ABSTRACT

Improvements in ion-sensors, calibration and mode of use are
discussed. Surfactant, calcium, nitrate and chloride ion-
selective electrodes illustrate improvements in ion-sensors. As
well as the sensors themselves, reference is made to grafting
ion-sensors to polymer matrices and to the use of bacteria in
membranes.

The use of ion-buffers, especially for calcium ions forms
the main plank of the discussion on calibration, and includes
reference to the MINIQUAD computer program for discerning complex
species in solution. The 'litre beaker' and 'spiking' methods
of calibration are also discussed.

In the final section, developments in mode of use of ion-
selective electrodes is illustrated with respect to micro electrodes, ISFETs, microprocessor based millivoltmeters and flow
injection analysis.

INTRODUCTION

About twenty years ago the composition of glasses was modified for selective response to cations other than hydrogen ions.
At this time, also, the advantages of a wider range of selectively sensitive electrodes were becoming appreciated, and in
1961 Pungor and Holós-Rokosinyi[1] described the selective working
of precipitate-based ion-sensitive electrodes. This and the
commercial availability of the fluoride ion-selective electrode
in 1966 heralded the resurgence of interest in analytical potentiometry to the extent that there is now a large bank of selective ion-sensors which can be used over wide concentration and
activity ranges in various branches of science.

As we look back it is possible to highlight stages in the
development and use of ion-selective electrodes and focus on
precipitate-based sensors, crystal membrane sensors, liquid
ion-exchanger sensors - both as liquid membranes and as membranes
trapped in poly(vinyl chloride) (PVC), neutral and charged carrier
sensors - also as liquid and PVC trapped membranes, applications

123

in divers fields including enzyme and substrate sensing, their
use in micro electrodes, and their use by Gran's plot and
related procedures.

From time to time, developments may be reviewed during
plenary lectures at Symposia or in articles, such as those on
micro ion-selective electrodes for intracellular ions, progress
in designing calcium ion-selective electrodes, ion-sensitive
field effect transistors, use of known addition/Gran's plot and
related methods with ion-selective electrodes, ion-sensitive
electrode screening tests in cystic fibrosis, application of
ion-selective electrodes in water analysis, etc. in the twice
yearly *Ion-Selective Electrode Reviews*.[2] Here, it is proposed
to highlight some of the improvements in ion-selective electrodes in terms of improved ion-sensors, improvements in calibration and improvements in the mode of use which can lead to
better application of ion-selective electrodes.

IMPROVED ION-SENSORS

Really outstanding ion-selective electrodes are few in
number and new electrodes for fresh ions rarely match those of
fluoride, sulphide, calcium and potassium in quality. Nevertheless, electrodes for use in difficult areas of analysis can
easily find a niche and flourish. For example, electrodes responding to surfactants may be based on membranes containing
an ion-association complex, such as cetyltrimethylammonium
dodecylsulphate in 1,2-dichlorobenzene.[3] They are an attractive alternative to dye transfer methods of surfactant analysis,
being useful as direct sensors and as indicator electrodes in
potentiometric titrations. However, because the electrodes
exhibit sharp breaks in e.m.f./log[surfactant] plots at the
critical micelle concentration (cmc), they are less useful in
the direct mode, although the break provides an alternative
means of determining the cmc.

Because of the nature of the analytical problems with surfactants, interest in trying to improve the quality of the
sensors continues, and the addition of a quaternary ammonium
sulphonate salt to the nylon matrix of an electrode responding
to sulphonate ions improves electrode response to pentadecyl-
benzenesulphonate[4] by giving a Nernstian response from the cmc
down to 10^{-7}M. Robustnuss is another desirable feature in this
area, as indeed in all areas where ion-selective electrodes are
employed. Thus. while trapping the surfactant sensor in PVC
yields stronger membranes than the supported liquid systems.
actual grafting of the sensor on to the polymer matrix can lead
to greater improvements. Thus, cationic surfactant electrodes
have been made by sulphonating PVC and converting to the tetraalkylammonium form.[5] Anionic surfactant electrodes may be
made by grafting tetraalkylammonium bromide to PVC and converting to the alkylsulphate or alkylbenzenesulphonate form as
appropriate.[5]

124

Grafting of sensors to the polymer matrix has been attempted for other electrodes types. Alkyl hydrogen phosphate has been grafted to the hydroxyl groups of a vinyl alcohol/vinyl chloride copolymer.[6] The calcium ion-selective electrodes obtained were similar in performance and general behaviour to those made by merely trapping of Orion 92-20-02 liquid ion exchanger in PVC. However, there was no gain in electrode lifetime to compensate for the extra effort of grafting the sensor to the matrix. Nevertheless, this approach is being continued in further studies on the grafting of sensor and solvent mediator.[7]

An alternative grafting approach for calcium ion sensors involves cross-linking a styrene-butadiene-styrene tri-block copolymer with triallyl phosphate and hydrolysing.[8] Unfortunately, this electrode shows poor calcium over sodium selectivity, although the addition of a solvent mediator, such as dioctyl phenylphosphonate may well promote selectivity favouring calcium in the same way as it does for non-grafted dialkylphosphate sensors.[9]

Although robustness and long operational lifetimes are very desirable features for ion-selective electrodes, selectivity and sensitivity take priority in many analytical applications. This is especially true for calcium ion-selective electrodes in studies such as detergent washing and physiological systems. For these areas, the newer calcium sensors based on calcium bis di(4-octylphenyl)phosphate and neutral carrier sensors have crossed boundaries in scope by their considerably improved selectivity towards calcium over sodium, potassium and magnesium.

Calcium bis di(4-octylphenyl)phosphate used in conjunction with di-n-octyl phenylphosphonate[10-12] or tri-n-pentyl phosphate[13] as solvent mediator has superseded calcium bis dialkylphosphates as an effective calcium ion-sensor and it matters but little in terms of calcium ion-selective electrode response whether the octyl group is the n-octyl chain or the isomeric 1,1,3,3-tetramethylbutyl.[12-15] This is important, since the starting material in the synthesis of the sensor is the 4-alkyl phenol and the 1,1,3,3-tetramethylbutyl isomer is much more readily available.[14-16]

The neutral carrier sensor, N,N'-di[(11-ethoxycarbonyl) undecyl]-N,N'-4,5-tetramethyl-3,6-dioxaoctane diamide,[17-19] is also a very effective calcium ion-sensor, and despite its more difficult synthesis than the di(4-octyl)phosphates it possesses an edge by its rather better selectivity for calcium over magnesium.[20] This neutral carrier is in a short-list[21] of five of the most attractive synthetic carrier molecules for sensing metal cations, the other four being selective for Li^+, Na^+ and Ba^{2+} as appropriate.

Nitrate ion-selective electrodes are widely used in water analysis, so that there has been an incentive for improvements here. As might be expected from development work on the PVC matrix membrane barium ion-selective electrode,[22] the use of a high viscosity solvent mediator (2-nitrophenyl octyl ether or

2-nitrophenyl phenyl ether) with tris(bathophenanthroline)
nickel(II) nitrate gives PVC matrix membrane electrodes of
longer lifetimes than when the less viscous 2-nitro-4-cymene is
employed.[23] There is also an extension in the linear calibration range to 10^{-5}M nitrate as has also been observed by using
tetra-alkylammonium nitrate sensors of alkyl chains with C_8 or
longer in conjunction with dibutylphthalate as solvent mediator
in PVC.[24] Indeed, the calibration range can even extend to
pNO_3 = 5.5 for tetradodecylammonium nitrate.[24,25]

A novel area in the field of selective potentiometric ion-
sensors is the development of bacterial membrane electrodes.
This may lead to specificity of the kind not previously realised,
although to date the development progress is not outstanding.
The devices depend on placing a paste of living bacteria between
say, the gas permeable membrane of an ammonia gas sensing
electrode and a dialysis membrane. In this way electrodes for
L-aspartate *(Bacterium cadaveris)*)[26] nitrate *(Azobacter vine-
landii)*[27] etc. have been prepared.

A bacterial membrane electrode for nitrate[27] had a life-
time of about 2 weeks following conditioning in a buffer solution. When not in use it was stored in the growth medium in
order to ensure reproducible day-to-day response. The response
to nitrate was rectilinear in the range 10^{-5} to 8×10^{-4}M
nitrate, but there was interference by nitrite,hydroxylamine,
copper(II), mercury(II), chlorate and urea.[27] Chloride, sulphate, sulphite and some amino acids could be tolerated.[27]

The constructional principle of the bacterial membrane
electrodes is similar to that of enzyme electrodes,[28] but as
with these the wider use of bacterial membrane electrodes can
well be hindered by the lack of widely available commercial
'self-contained' electrodes. It should not be beyond scientific
ingenuity and commercial enterprise for lines of individual
enzyme and bacterial electrodes to be available on the market.

Electrodes with mixed mercury(I) chloride/mercury(II) sulphide membranes[29,30] have proved superior to silver/silver
chloride for determining low level chloride in boiler water[30]
at concentrations of 0.01-1.0 µg cm^{-3}. At 0.1 µg cm^{-3} of
chloride, better precision was obtained with the mercury(II)
chloride/mercury(II) sulphide impregnated selectrode style of
electrode than with the silver/silver chloride type (0.004 µg
cm^{-3} compared with 0.04 µg cm^{-3} for the total standard deviation), although at 1 µg cm^{-3} the two electrodes were equally
precise.[30] At concentrations below 1 µg cm^{-3} the mercury(I)
chloride electrode is much more sensitive than the silver chloride type, for example, the difference between the e.m.fs observed
in 1.0 and 0.1 µg cm^{-3} chloride solutions is 51 mV for the
mercury(I) chloride electrode and 19 mV for silver chloride
electrodes.[30] Between 0.1 and 0.01 µg cm^{-3} the corresponding
differences were 13.5 and 2 mV, respectively.

The mercury(I) chloride electrode is, however, slower in reaching equilibrium than the silver chloride type, taking 5 min to reach a steady e.m.f. while the latter requires less than 1 min.[30] Also, with the mercury(I) chloride electrode the best results are obtained with the impregnated graphite Selectrode,[30] for the precision of analysis with pelleted membranes[31] is poorer.

IMPROVEMENTS IN CALIBRATION

The use of fluoride ion buffers consisting of fluoride complexes of H^+, Zr^{4+}, Th^{4+} and La^{3+} to study[32] the behaviour of the fluoride electrode at very low fluoride ion levels showed that linear response held right down to $10^{-9.5}M$ fluoride and not just to $10^{-6}M$ as deduced from serial dilution. This was an important concept in extending the calibration range of ion-selective electrodes.

Buffer systems containing various proportions of copper(II) ions to EDTA or NTA were used for calibrating copper electrodes[33] The ionic strength was maintained at about 0.1M in order to keep constant the activity coefficient of the free Cu^{2+} ions in equilibrium with the Cu(II)/EDTA or Cu(II)/NTA complex. pCu values for the different buffers were calculated[33] from published stability constants obtained by Ringbom's method.[34]

Buffers have been formulated for calcium ions,[10,35,36] cadmium ions,[37] and lead(II) ions,[38] and used for electrode calibration.[10,35-38] The original principles applying to calcium ion-selective electrodes[10] have been adopted and tested by many workers,[11,19,35,36,39-41] and the electrodes can be calibrated to ca $10^{-8}M$ free calcium ions.[15] It is significant that the calcium ion buffer calibration extrapolates to that for serial dilution of calcium ion standards.[35] This supports the view[41] that "the calcium electrode is essential for reliable determination of the calcium ion concentration in complex buffers" as deduced from the almost identical pCa values obtained by calcium ion-selective electrode (Radiometer F2112 calcium Selectrode based on a di(4-octylphenyl)phosphate sensor) and those calculated from stability constants of calcium/EGTA and calcium/NTA systems. Good agreement was obtained even in the presence of magnesium and adenosine triphosphate.[41]

Long range calibrations of electrodes with calcium bis di(4-n-octylphenyl)phosphate sensor have been tested for calcium/anion ligand systems in the range of $<10^{-7}$ to $>10^{-3}M$ free calcium ions.[40] The data obtained were used to calculate conditional stability constants for various calcium/ligand systems in 0.1M sodium chloride[40] and there was good agreement with literature data as summarised in Table I.

With certain systems, especially those involving tripolyphosphate and pyrophosphate it is necessary to consider the possible formation of species other than 1:1 complexes, for example, protonated complexes, the relevant equilibria being

$$HP_3O_{10}^{4-} \rightleftharpoons H^+ + P_3O_{10}^{5-} \quad \text{and} \quad H_2P_3O_{10}^{3-} \rightleftharpoons 2H^+ + P_3O_{10}^{5-}$$

for tripolyphosphate and

$$HP_2O_7^{3-} \rightleftharpoons H^+ + P_2O_7^{4-} \quad \text{and} \quad H_2P_2O_7^{2-} \rightleftharpoons 2H^+ + P_2O_7^{4-}$$

for pyrophosphate. The possibility of these complexes arise from the pH of the titration test solutions and their handling required the use of a computer program capable of computing formation constants and species distribution in equilibrium systems. For this work MINIQUAD[42] was selected for its capability of dealing with potentiometric data, but the program required some modification.[40]

Since the MINIQUAD program is sensitive to small changes in the constants for the above equilibria, the constants were determined under the conditions of measurements with calcium by straightforward acid-base titration.[40] Thus, typically, sufficient 1M hydrochloric acid (ca 0.2 cm^3) was added to 100 cm^3 of 10^{-3}M of the appropriate phosphate solution to adjust the pH to about 4, and the ionic strength was adjusted to 0.10M with sodium chloride. Sodium hydroxide solution (0.1M) was added from an Agla burette until a pH of 9.5-10 was reached and the monitored pH obtained during the addition used as input to MINIQUAD together with titrant volume.[40]

For the calcium ion-selective electrode measurements with both of the phosphate systems, 100 cm^3 calcium chloride solutions (5 x 10^{-5} to 10^{-4}M and I = 0.10M) were titrated at 25°C with freshly prepared 0.10M sodium tripolyphosphate (or pyrophosphate) solution from an Agla burette to a final phosphate concentration of 5 x 10^{-4}M. Prior to the titration the pH of each solution was adjusted to between 8 and 9. After each addition, the volume added, calcium ion-selective electrode potential and pH were recorded for use as MINIQUAD input data.[40] The results showed that the most probable species for the system are those named in Table II.

Calcium ion-selective electrodes have also been put to the test by measurements of low levels of calcium ions in physiological systems. Thus, measurements with microelectrodes[43] have yielded data of 7 x 10^{-7}M calcium in *Aplysia* giant cells[44] and 4.5 x 10^{-7}M in snail giant neurons[45] with di[4-(1,1,3,3-tetramethylbutyl)phenyl]phosphate sensor, and ca 10^{-7}M calcium for *Balanus nubilus* muscle fibres using the ETH 1001 carrier in a 20 μm Ca^{2+} electrode.[46]

Improvements in the technique of calibration add to the degree of confidence with which ion-selective electrodes can be used. In this respect it frequently happens that as a result of better calibration electrodes are linear to beyond what was previously thought of as the detection limit. An early method for confirming longer calibration ranges was the 'litre beaker' method[47] whereby a litre (1 dm^3) of sample background containing an ion-selective and reference electrode pair is kept stirred and has added to it small increments of a concentrated solution of the primary ion, A. The resulting concentrations and cell potentials are matched for each addition.

The principle of the 'litre beaker' method may be used on a reduced scale by using an Agla micrometer syringe burette for "spiking" small increments of fresh concentrated standardising solution into the background solvent or solution matrix.[48] In this way, a solid-state silver sulphide membrane ion-selective electrode has been calibrated with standards down to ca 10^{-8}M silver and 2×10^{-7}M sulphide ions, respectively.[48] The electrode will respond to even lower levels of silver and sulphide, but calibration is bedevilled by sorption of silver on vessel walls and of oxidation of sulphide by spurious traces of oxygen-even in the presence of anti-oxidants. Slow electrode kinetics at low concentrations/activities also mean that long times are required for equilibrium e.m.fs. to be reached in the circumstances.

As an alternative to spiking, standardising solution, A, may be brought into the calibration system by coulometric generation by electrodes immersed in the solution, as has been used for calibrating iodide, and silver ion-selective electrodes.[49] The advantage of this and the spiking systems mentioned above is that full advantage is taken of the faster and better electrode response that emanates from use in dynamic conditions.

DEVELOPMENTS IN MODE OF USE

The use of ion-selective electrodes under the best operating conditions is imperative. Thus, with the fluoride electrode, pH, complexation and ionic strength interferences can be smoothed out by the use of TISAB (total ionic strength adjustment buffer). For calcium ion-selective electrodes with dialkylphosphate sensors it is beneficial to keep pH under control in order to avoid fluctuating e.m.fs at below pH 5.5. Apart from solution conditions, there are many other developments in the mode of use of ion-selective electrodes that amount to definite improvements.

In micro electrode terms, it can be beneficial to have the ion-selective electrode and reference electrode as a single unit. This has been achieved by using double-barrelled electrodes of which several are mentioned by Brown and Owen in their recent review on micro-ion-selective electrodes.[43]

Also significant are developments in ISFETs (ion-selective field-effect transistors).[50] These can be looked upon as an evolution of measurement with ion-selective electrodes whereby during e.m.f. measurement in conjunction with a reference electrode, the ion-selective electrode is connected by a wire to an FET input of a millivoltmeter. In the next stage of evolution, the lead connecting the ion-selective electrode is made shorter until it is completely disposed of, and the ion-selective electroactive membrane is placed directly on the insulator of the FET and the input transistor is placed in the measured solution.

In this stage the integration of the ion-sensitive membrane and the solid state amplifier is complete, but the resulting ISFET is in practice very, very small.[50]

ISFETs have the advantage of transforming the high resistance of ion-selective membranes into low output impedance with the result that the signal probe does not have to be heavily shielded in order to minimize electrical interferences. This can make ion-selective probes much more amenable to use in many biomedical applications and complement the use of micro ion-selective electrodes, but there is the constraint that while ion-selective electrodes can be made and studied under very modest conditions, ISFETs can be fabricated only in a very well equipped and relatively expensive semiconductor laboratory.[50]

In these days of great strides in the incorporation of microprocessors into analytical instrumentation, it would not be appropriate to conclude without reference to some of the features being made available in the field of ion-selective electrodes. Standard addition methods are extremely helpful in applications of ion-selective electrodes,[51] but the antilogging of the standard addition equation is cumbersome and needs the convenience of a calculator or similar device. However, the developments in microprocessors have led to E.M.F. Measuring equipment that copes not only with the mathematical operations, but also with different electrode slopes and concentrations of standards. For example, the Orion model 901 meter has a high impedance amplifier and analog-to-digital converter whose output is fed to a microprocessor pre-programmed with the appropriate equations for calculating pIon and concentration.[52] For a solution of unknown concentration the program within the Orion 901 meter is based on providing a set of input data, namely, the cell potential for the unknown solution. Concentration of standard, electrode slope, and/or blank correction and MODE (KA or known addition in the present context) are stored in the memory and permit the microprocessor to solve the standard addition equations.[52] Other MODES are available in the instrument, for example, Analyte Addition (AA) and Analyte Subtraction.

By using ordinary sample by sample techniques up to about 20 to 30 samples per hour may be analysed manually with ion-selective electrodes, but if the devices are incorporated in certain flow injection systems[53] this number can be greatly increased and may exceed 80 samples per hour as described for potassium and nitrate ion-selective electrodes for measuring potassium and nitrate in fertilisers.[54,55] Such a non-segmented flowing system where the response is a fraction of the total depends on freedom from sample variations, such as, viscosity and surface tension. In such circumstances it may be better to employ alternative flow-through methods.[56] Nevertheless, the general role of flow methods is to provide rapid analysis and they provide the dynamic conditions under which ion-sensors best perform in terms of more rapid response times.

REFERENCES

1. E.Pungor and E.Holós-Rokosinyi, Acta Chim.Acad.Sci.Hung., 27, 63 (1961).
2. J.D.R.Thomas (Editor) Ion-Selective Electrode Reviews, Pergamon Press, Oxford, Volume 1 (1979), Volume 2 (1980).
3. B.J.Birch and D.E.Clarke, Analytica Chim.Acta, 67, 387 (1973).
4. S.H.Hoke, A.G.Collins and C.A.Reynolds, Anal.Chem., 51, 859 (1977).
5. S.Cutler, P.Meares and G.Hall, J.Electroanalyt.Chem., 85, 145 (1977).
6. L.Keil, G.J.Moody and J.D.R.Thomas, Analyst, 102, 274(1977).
7. P.C Hobby, G.J.Moody and J.D.R.Thomas, to be published.
8. L Ebdon. A.T.Ellis and G.C.Corfield, Analyst, 104, 730 (1979).
9. A.Craggs, L.Keil, G.J.Moody and J.D.R.Thomas, Talanta, 22, 907 (1975).
10. J.Růžička, E.H.Hansen and J.C.Tjell, Analytica Chim.Acta, 67, 155 (1973).
11. H.M.Brown, J.P.Pemberton and J.D.Owen, ibid, 85, 261(1976).
12. L.Keil, G.J.Moody and J.D.R.Thomas, ibid, 96, 171 (1978).
13. G.J.Moody, N.S.Nassory and J.D.R.Thomas, Analyst, 103, 68 (1978).
14. J.D.R.Thomas, Lab.Practice, 27, 857 (1978).
15. G.J.Moody and J.D.R.Thomas, Ion-Selective Electrode Revs., 1, 3 (1979).
16. A.Craggs, P.G.Delduca, L.Keil, B.J.Key, G.J.Moody and J.D.R.Thomas, J.Inorg.and Nucl.Chem., 40, 1483 (1978).
17. D.Ammann, E.Pretsch and W.Simon, Helv.Chim.Acta, 56, 1780 (1973).
18. D.Ammann, R.Bissig, M.Güggi, E.Pretsch, W.Simon, I.J.Borowitz and L.Weiss, ibid, 58, 1535 (1975).
19. D.Ammann, M.Güggi, E.Pretsch and W.Simon, Anal.Letters, 8, 709 (1975).
20. T.S.Tsieng, private communication.
21. W.E.Morf and W.Simon, Hungarian Sci.Instruments, 41, 1 (1977).
22. A.M.Y.Jaber, G.J.Moody and J.D.R.Thomas, Analyst, 101, 179 (1976).
23. A.Hulanicki, Magdalena Maj-Zurawaska and R.Lewandowski, Analytica Chim.Acta, 98, 151 (1978).
24. H.J.Neilsen and L.H.Hansen, ibid, 86, 1 (1976).
25. Judith A.Wright and P.L.Bailey, in E.Pungor (Editor) Ion-Selective Electrodes, Akadémiai Kiadó, Budapest, (1978), p.603.
26. R.K.Kobos and G.A.Rechnitz, Anal.Letters, 10, 751 (1977).
27. R.K.Kobos, D.J.Rice and D.S.Flournoy, Anal.Chem., 51, 1122 (1979).
28. G.G.Guilbault, in G.Svehla (Editor) Wilson and Wilson's Comprehensive Analytical Chemistry, Elsevier, Amsterdam, Volume VIII (1977).
29. J.F.Lechner and I.Sekerka, J.Electroanalyt.Chem., 57, 317 (1974).
30. G.B.Marshall and D.Midgley, Analyst, 103, 438 (1978).

31. I.Sekerka, J.F.Lechner and R.Wales, Water Res., 9, 663 (1975).
32. Elizabeth W.Baumann, Analytica Chim.Acta, 54, 189 (1971).
33. R.Blum and H.M.Fog, J.Electroanal.Chem., 34, 485 (1972).
34. A.Ringbom, Complexation in Analytical Chemistry, Interscience, New York (1963).
35. A.Craggs, G.J.Moody and J.D.R.Thomas, Analyst, 104, 412 (1979).
36. B.J.Birch, A.Craggs, G.J.Moody and J.D.R.Thomas, in E.Pungor (Editor), Ion-Selective Electrodes, Akadémiai Kiadó, Budapest (1978), p.335.
37. J.Růžička, E.H.Hansen and J.C.Tjell, Analytica Chim Acta, 63, 115 (1973).
38. E.H.Hansen and J.Růžička, ibid, 72, 365 (1974).
39. B.J.Birch, A.Craggs, G.J.Moody and J.D.R.Thomas, J.Chem. Educ., 55, 740 (1978).
40. A.Craggs, G.J.Moody and J.D.R.Thomas, Analyst, 104, 961 (1979).
41. O.Scharff, Analytica Chim.Acta, 109, 291 (1979).
42. P.Gans, A.Sabatini and A.Vacca, Talanta, 21, 53 (1974).
43. H.M.Brown and J.D.Owen, Ion-Selective Electrode Reviews, 1, 145 (1979).
44. J.D.Owen, H.M.Brown and J.P.Pemberton, Analytica Chim. Acta, 90, 241 (1977).
45. G.R.J.Christofferson and L.Simonsen, Acta physiol.Scand., 101, 492 (1977).
46. C.C.Ashley, T.J.Rink and R.Y.Tsien, J.Physiol., 256, 27P (1978).
47. Orion Research Inc., Newsletter, 2, 42 (1970).
48. D.J.Crombie, G.J.Moody and J.D.R.Thomas, Analytica Chim. Acta, 80, 1 (1975).
49. P.L.Bailey and E.Pungor, ibid, 64, 423 (1973).
50. J.Janata and R.J.Huber, Ion-Selective Electrode Reviews, 1, 31 (1979).
51. M.Mascini, Ion-Selective Electrode Reviews, 2, (1980).
52. G.J.Moody and J.D.R.Thomas, Lab.Practice, 28, 125 (1979).
53. J.Růžička and E.H.Hansen, Analytica Chim.Acta, 78, 145 (1975).
54. E.H.Hansen, A.K.Ghose and J.Růžička, Analyst, 102, 705 (1977).
55. E.H.Hansen, F.J.Krug, A.K.Ghose and J.Růžička, ibid, 102, 714 (1977).
56. E.Pungor, G.Nagy, Zs.Fehér and K.Tóth, Proc.Anal.Div.Chem. Soc.,16, 347 (1979).

Table I. Stability constants for calcium-ligand systems from calcium ion-selective electrode measurements at 0.1M ionic strength in 0.1M sodium chloride[17]

Complex	log β [a]	Literature log β
CaCitrate	3.42	3.17 to 4.90
CaMalonate	1.52	1.46 to 2.49
CaMalate	2.00	1.80 to 2.66
CaOxalate	2.54	1.66 and 3.00
$[CaEDTA]^{2-}$	10.93	10.42 to 11.0
$[CaNTA]^{-}$	6.31	6.46 and 6.57
Ca,SO_4	1.39	2.0 to 2.48
$CaHPO_4$	1.87	1.70
$[CaP_3O_{10}]^{3-}$	5.05	5.20 and 6.41
$[CaP_2O_7]^{2-}$	4.33	5.39

a The range of $[Ca^{2+}]_{free}$ levels measured was 3.3×10^{-8}M to 4.8×10^{-3}M

Table II. Stability constant (log β) data for calcium tri-polyphosphate and calcium pyrophosphate systems obtained at 0.1M ionic strength in 0.1M sodium chloride using a computer program (MINIQUAD)[40]

Equilibrium at 25°C	Titration points (n)	Log β (σ_{n-1}) [a]
$Ca^2 + P_3O_{10}^{5-} \rightleftharpoons [CaP_3O_{10}]^{3-}$		5.05 (0.02)
$Ca^{2+} + 2(P_3O_{10})^{5-} \rightleftharpoons [Ca(P_3O_{10})_2]^{8-}$	50	9.41 (0.07)
$Ca^{2+} + P_2O_7^{4-} \rightleftharpoons [CaP_2O_7]^{2-}$		4.33 (0.01)
$Ca^{2+} + 2(P_2O_7)^{4-} \rightleftharpoons [Ca(P_2O_7)_2]^{6-}$	91	7.21 (0.08)

a See Reference 40 for details of computer model

QUESTION

Participants of the discussion: R.G.Bates, J.D.R.Thomas

Question:
How did you determine the stability constants with EDTA and
NTA; what were the ionic strength and the pH ? Was there any
indication of complexation with the buffer itself ?

Answer:
I would like to refer to the actual reference of this work:
Analyst, 1979, 104, p. 961. We have not considered any complexation between the complexing agents and the sodium in the
buffer.

PROBLEMS RELATED TO THE INTERPRETATION OF RESPONSE TIME CURVES OF ION-SELECTIVE ELECTRODES

K. TÓTH, E. LINDNER AND E. PUNGOR

Institute for General and Analytical Chemistry, Technical University, Budapest, Hungary

INTRODUCTION

In the last few years the interest of scientists has turned again to response time studies, in spite of the known-difficulties and problems related to this type of work. This may be explained by the wide-spread application of continuous analytical systems employing electroanalytical sensors and by the renewed interest towards obtaining kinetic information about the electrode response.

In our laboratory a great deal of interest has been shown in studying dynamic response of ion-selective electrodes with the activity step method since the early stage of ion-selective electrode research /1,2/.

Naturally the dynamic characteristics of an ion-selective electrode can be studied by different techniques e.g. current pulse, or potential pulse method or the activity step method. In our opinion the activity step method is the most suitable for this type of work since it is realized under zero current membrane condition and therefore, it has been used throughout our work. Although small activity steps are theoretically favoured /3/, its practical realization especially at very high flow-rates encounters several difficulties arising mainly from the measuring technique itself /the change of the diffusion and streaming potentials etc./.

In spite of the fact that we have employed a special device /4/ ensuring an almost ideal activity step, and a very high flow-rate, the time constants of the response time curves

corresponded to film diffusion values as the concentration of
the stagnant film of initial solution changes by diffusion
migration to the new stepped value. Following from all this,
only those processes show up separately on a response time
curve that are consistent either with diffusion on the solution
side of the interface or slow surface processes. Accordingly
our attention has been directed to find conditions to eliminate
the overlapping effect of film diffusion. This has been met
especially in range of the lower detection limit of the electrode where the exchange current density is decreased significantly, as well as in the two ion range.

RESULTS AND DISCUSSION

Firstly the response characteristics of an ion-selectivly
electrode in the range of the lower detection limit has been
studied. The way, how the activity step method has been carried
out is shown in Fig.1. The response time curves recorded following a one decade activity step at different primary-ion
concentration levels have been compared with the help of
normalized initial slope values /4/. The corresponding results
are summarized in Fig. 2, for activity increase and decrease
at different flow-rates. On the basis of this it can be stated
clearly as has been published earlier that:
- the electrode response is faster at activity increase than
at activity decrease,
- the response time is shorter at higher flow rates in a given
activity range. Our latest results, however, show clearly that
there exists a critical activity value bellow that the rate
of response is flow-rate independent at both directions of
activity step changes. This suggests that the diffusion model
and the corresponding mathematical equation /5/ describes the
response time curve only in the flow-rate dependent range
where the corrected initial slopes are concentration level
independent. However, recording response time curves bellow
the critical activity level, it is believed that the response
time data correspond to processes following the film diffusion/6/

In the second part of this paper the dynamic behaviour of the iodide ion-selective electrode in two ionic range is presented. The activity steps, which can be used for this type of study, are shown in Fig. 3 with the help of a hypothetic calibration graph.

Generally in the two ion response range the response time curves can be recorded by ensuring a constant level of interfering ion and introducing a primary ion activity jump, or on the other way round. However, if the first possibility is chosen then the overlapping effect of film-diffusion considering a fast surface exchange reaction make same-times the evaluation more difficult. This type of work was carried out by Fleet et al. /7/. If one works with constant primary ion level and use an interfering ion activity step than one can record a non-monotonically changing potential response, which are non-equilibrium effects. These, however have not yet been sufficiently characterized.

In our work the measurements were carried out under conditions that the equilibrium bromide ion interference was smaller than 18 mV. The selection of this concentration ratio was justified partly by practical application points of view and partly by the fact that according to earlier publications the largest overshoot was obtained under the condition where the electrode just starts to deviate from the Nernstian response /8/. Employing an interfering activity step the magnitude of the transient signal is much larger than that of the equilibrium potential difference calculated, and the rate of its relaxation is relatively slow, thus, it can easily be studied in the case of precipitate based electrodes as well.

Only a few papers have been published in the literature on this subject /8-12/ and these are dealing only with glass electrodes and other liquid ion-exchanger electrodes. Camman /13/ has explained the phenomenon with the help of an exchange kinetic model, while Morf /14/ has used a diffusion model considering a segmented membrane with different selectivity factors with the assumption that ions of the same charge have the same mobilities within the membrane phase. Others /15/, however, explained the phenomenon with the different mobilities

of the primary and interfering ions within the membrane phase, and diffusional relaxation of the transient signal directed towards the membrane. In our work the dynamic response of an iodide electrode has been studied in the two ion range employing an interfering ion i.e. bromide ion activity step as function of various experimental parameters, such as
- flow rate /Fig. 4/
- the level of the interfering ion /the extent of interference/ /Fig. 5/.

It was found that the first parts of the transient signal following a bromide increase or decrease is extremely fast and flow-rate dependent /Fig. 4/. Its time constant is in the ms-range and the shape of this running up part is very similar to the first part of response time curves obtained at a primary ion activity step recorded in solutions containing primary ions only. On the basis of this it can be supposed that the processes resulting the first part of the transient signals in the two ion range are controlled by film-diffusion. The second, the so-called relaxation part of the transient signal having a several orders of magnitude larger time constant than the first one, is also flow-rate dependent. In connection with the level of the interference one can say that the magnitude of the transient signal is interfering ion level dependent and it is almost proportional to the logarithm of the interfering ion activity /Fig. 5/. Studying the transient signal in the two ion range for a longer period of time a third very slow section shows up on the transient signal. This phenomenon partly can be attributed to a slow surface reaction /16/.

On the basis of the experimental findings described a qualitative model can be offered for the interpretation of transient behaviour of the iodide ion-selective electrodes in the two ion range. Accordingly the transient signal is supposed to be a result of the following processes, which take place consecutively or parallely to each other:
- the diffusion of the interfering ion through the stagnant solution film from the bulk of the solution to the electrode surface or in the opposite direction;
- fast interfacial processes /bromide ion adsorption, ion-

-exchange process, iodide desorption/;

- iodide ion diffusion from or towards the electrode membrane;

- slow surface processes resulting in the change of the surface composition and morphology and the connected diffusion processes.

The first two processes supposed to be responsible for the first running-up section of transient signal, while the third is to the relaxion towards the solution. The last process results in a drift-like slow surface modification.

With the model offered the flow-rate dependence of the different parts of the transient signal can easily be interpreted. Accordingly the rate of the running-up section of the transient signal is determined by the diffusion gradient of the interfering ion assuming that the rate of exchange kinetics is faster than that of diffusion. The magnitude of the signal however, is proportional to the amount of iodide ion generated in given period of time, in a given solution volume by the interfering ion. Both parameters the rate of running up and the magnitude of the overshoot are greatly effected by the thickness of stagnant diffusion film. The direction of the potential jump, however, according to our model will correspond to an iodide ion activity increase at the electrode surface as a result of an interfering ion activity increase.

The second part of the transient signal "the potential decay" is explained by the diffusion relaxation of the primary ion - generated in a surface exchange reaction - towards the solution. Thus, as expected the direction of potential change corresponds to the decrease of the primary ion. Due to the small diffusion gradient, the time constant of this second diffusion controlled relaxation process, obviously much larger than that of the running up section.

The transient signal recorded at interfering ion activity decrease can similarly be explained.

The third slowest part of the transient signals is explained to be connected to slow surface modification as it was observed that at the use of the electrode in the two ionic solution the E_o and K_{ij} -values were slowly changed.

CONCLUSION

In conclusion it can be stated that our model which is based
on the assumption that the overshoot of the transient signal
disappears in a diffusion relaxation towards the bulk of the
solution - found to be appropriate to interpret the transient
characteristics of the iodide electrode in the two ion range.
According to this model, it was not necessary to suppose the
existence of an outer membrane layer having a selectivity
factor different from the bulk of the membrane. Moreover, non-
-expected transient signals were found with an iodide electrode
having a thin AgBr layer at its surface in the two ion range
/16/.

The model is found to be valid also if the mobilities of
the two ions are almost the some.

In our opinion our model can be extended to interpret the
transient signals of other type of ion-selective electrodes
also such as glass or other ion-exchanger based electrodes, in
the two ion range.

REFERENCES

1. K. Tóth, I. Gavallér and E. Pungor, Anal.Chim.Acta 57,
   /1971/ 131
2. K. Tóth and E. Pungor, Anal.Chim.Acta, 64, /1973/ 417
3. R.P. Buck, Theory and Principles of Membrane Electrodes
   /Chapter 1 in Ion-Selective Electrodes in Analytical
   Chemistry, H.Freiser, ed./, Plenum Press, New York, 1978
4. E. Lindner, K. Tóth and E. Pungor, Anal.Chem., 48, /1976/
   1071
5. W.E. Morf, E. Lindner and W. Simon, Anal.Chem., 47, /1975/
   1596
6. E. Lindner, K.Tóth and E. Pungor, in preparation
7. B. Fleet, T.H. Ryan and M.J.D. Brand, Anal.Chem., 46, /1974/
   12
8. B. Karlberg, J. of Electroanal. Chem., 42, /1973/ 115

9. G.A. Rechnitz and G.C. Kugler, Anal.Chem., <u>39</u>, /1967/ 1682

10. J. Bagg and R. Vinen, Anal.Chem., <u>44</u>, /1972/ 1773

11. R.E. Reinsfelder and F.A. Schultz, Anal.Chim.Acta, <u>65</u>, /1973/ 425

12. D.E. Mathis, F.S. Stover and R.P. Buck, J. of Membrane Science, <u>4</u>, /1979/ 395

13. K.Camman, Das Arbeiten mit ionenselektiven Electroden, Springer-Verlag, Berlin, Heidelberg, New York, 1977

14. W.E. Morf, Anal.Letters, <u>10</u>, /1977/ 87

15. A.A. Belijustin, I.V. Valova and I.S. Ivanovskaja, in Ion-Selective Electrodes, Ed. E. Pungor, Akadémiai Kiadó, Budapest, 1978

16. E. Lindner, K.Tóth and E. Pungor, in preparation.

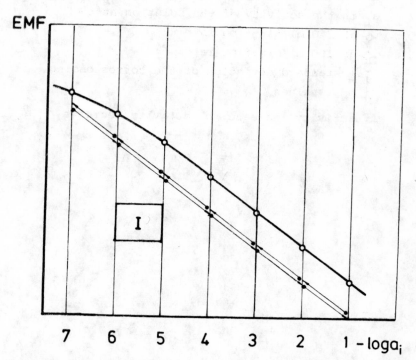

Fig. 1   A hypothetic calibration graph showing the activity range and activity steps employed for response time studies

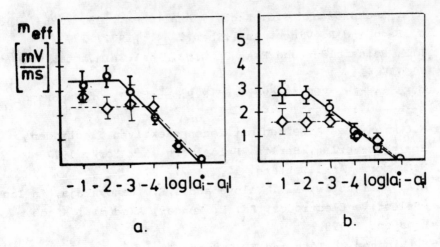

Fig. 2 Normalised initial slope values $/m_{eff}/$ versus
log $a_i^o - a_i$ graph corresponding to a one decade
activity step at different activity levels.
a_i^o is the activity of the solution at $t < 0$
a_i is the activity of the solution at $t > 0$
-o- at 140 ml/min flow rate
-◇- at 100 ml/min flow rate
$\pm$ standard deviation of the corresponding
m_{eff} values

a. activity increase; b. activity decrease

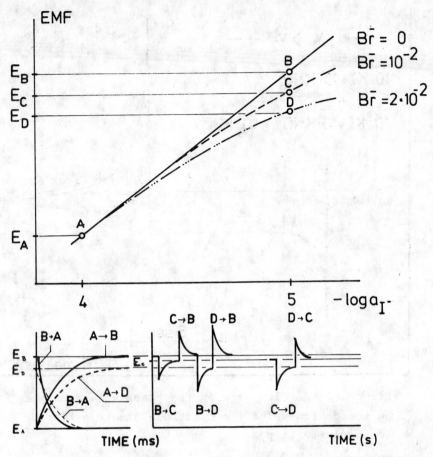

Fig. 3 Hypothetic calibration graph at different interfering
ion activity levels and the corresponding transient
signals obtained at various activity steps

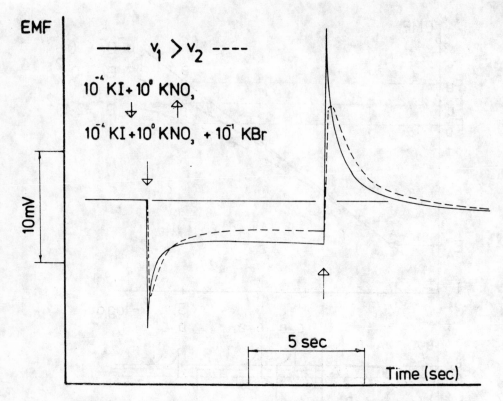

Fig. 4 Effect of the flow rate on the transient signal of
an iodide ion-selective electrode due to a bromide
ion activity step

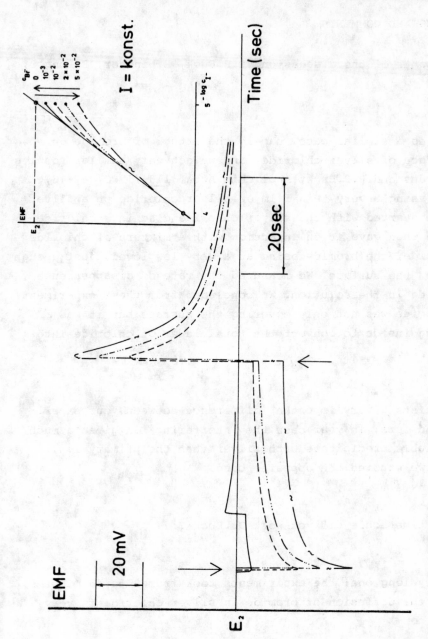

Fig. 5 Effect of the interfering ion activity level on the transient signal of an iodide ion-selective electrode due to a bromide ion activity step

145

QUESTIONS AND COMMENTS

Participants of the discussion: R.P.Buck, E.Lindner, E.Pungor,
K.Tóth

Comment:
We studied a similar case, namely the attack of bromide on
the surface of silver chloride, and we got very similar results
/R.K.Rhodes and R.P.Buck, Anal.Chim.Acta, 113 /1980/ 67-78/.
We were using a very thin film of silver chloride on a silver
disc. We started with the silver chloride disc in a chloride
solution then gave it on injection with a mixture of chloride
and bromide, the bromide being at a very low level, just enough
to attack the surface. We measured the rate of disappearence
of bromide in the solution. We concluded from these experiments
that bromide was not only covering the surface but it was
diffusing inside to convert the total layer of chloride into
bromide.

Answer:
There exists a very important difference between your experiments and ours: in your case the interfering ion gives a much
less soluble precipitate with silver than the primary ion,
whereas we studied the opposite case.

Question:
On what time-scale did you work Dr.Buck ?

Answer:
On a very long one: the experiments took from 5 to 45 minutes,
because the diffusion of bromide in silver chloride is very
slow.

Question:
Dr.Buck, how thick were the layers you used ?

Answer:
They were 2.4×10^{-4} cm thick.

ELECTROCHEMICAL STUDIES OF SOME SOLID-STATE ION-SELECTIVE ELECTRODES

YU. G. VLASOV

Chemistry Department, Leningrad University, Leningrad, 199004, USSR

CONTENT

147

1. INTRODUCTION

1.1. Classification of electrodes

Electrochemical investigation of ion-selective electrodes involves the consideration of a wide range of problems that depend on the conditions of measurements, on the membrane material and on the electrode design. In this connection it is worth considering briefly the general characteristics of electrodes including their classification, peculiarities of membranes and designs.

All the existing electrodes that are used for the determination of the concentration of substances in different media, can be devided into two large groups depending on the type of the particles being determined: ion-sensitive electrodes and electrodes sensitive to molecular products (Fig.1).

Under certain conditions the first group of electrodes allows to determine ion concentration in solutions and, in its turn, can be devided into two subgroups differing in principle: 1) electrodes of the 1st, 2nd kinds and electrodes on the basis of redox salt systems and 2) ion-selective electrodes (ISE). The principal difference between the 2nd kind electrodes and ion-selective electrodes based on the different mechanisms of their action (redox and ion-exchange, respectively) was shown in the work /1/ as a result of a detailed theoretical analysis of electrochemical phenomena in solution--membrane systems.

The second group of electrodes serves for the determination of the concentration of molecular products such as gases, organic molecules, etc., and it is represented by enzyme electrodes and electrodes sensitive to gases, for example.

Due to a number of useful properties (insensitiveness to redox potentials, great specificity to some ions, stability, etc.) differing them, in particular, from the electrodes of the first subgroup, ion-selective electrodes are widely used for practical determination of the concentration of ions in solutions both in the laboratory and in industry. Detailed description of the theory and application of ion-selective electrodes as well as the latest achievements in this field can be found in a number of monographs /2-13/ review papers by Buck /14-15/ and Nickolsky /16,17/ as well as in the special recently founded journal "Ion-Selective Electrodes".

In order to determine the role and importance of just solid-state ion-se-

148

lective electrodes that are dealt with in this paper, let us consider the classification of ion-selective electrodes. If ISEs are classified according to the principle based on the properties of the sensitive material used for ISE membranes, one can distinguish three different types of ISEs: 1) with liquid membranes, 2) with solid membranes and 3) ion-selective field effect transistors (ISFETs), the sensitive material of which is the combination of the above mentioned membranes with the field effect transistor(Fig.1). According to the same principle of classification solid ion-selective electrodes can be devided into homogeneous and heterogeneous; glass electrodes , electrodes with amorphous semiconducting (chalcogenide) membranes and single crystal membranes or membranes based on solid solutions being regarded as homogeneous while the electrodes the membranes of which are either pressed pellets made of a mixture of several polycrystalline salts or a mixture of these salts in a neutral matrix being regarded as heterogeneous.

At present solid membranes are used in most ISFETs as well, therefore, the classification of solid electrodes mentioned above is applied to ISFETs (see Fig.1), one more membrane type which is thin dielectric film (SiO_2 or Si_3N_4) being added to it. The classification shown in Fig.1 differs to some extent from the IUPAC classification /18/ by that it takes into account the new type of ion-selective electrodes - ISFETs, as well as by that pressed polycrystalline membranes based on mechanical mixture of salts have been transferred from the group of homogeneous one into that of heterogeneous.

1.2. Solid-state ion-selective electrodes

From the classification mentioned above one can see that solid ion-selective electrodes, including ISFETs, are the greater part of all ion-selective electrodes. From the practical point of view solid ISEs have a number of advantages over other kinds of ISEs, they are mechanical stability, relative chemical inertness, thermal stability, etc. Some solid ISEs with oxide membranes can be used at high temperatures, including melts. Electrodes of this type will not be considered in this paper because our aim is only electrodes used at room temperature. Being a particular field of research, glass electrodes also are not considered in this paper.

As it has been said already, the problem of electrochemistry of solid ion-selective electrodes is connected with both the material of membranes and with the design of electrodes. Fig.2 shows different schemes of const-

ruction of ion-selective electrodes. The electrode, without internal solution that is called an all-solid-state electrode (Fig.2), can be of the
greatest practical importance among general ion-selective electrodes. Such a
design results in some additional theoretical problems, that must be solved
to make the right choice of a membrane and a contact.

If one speaks about the material of membranes of ion-selective electrodes
(except glass electrodes), they can be divided into five groups:

1) single crystal membranes, e.g., on the basis AgCl, AgBr, $LaF_3:Eu^{2+}$,
   $LaF_3:Ca^{2+}$;

2) membranes on the basis of certain pure polycrystalline salts, e.g.Ag_2S;

3) membranes being a mechanical mixture or solid solutions of several
   salts, e.g.: $AgX-Ag_2S$, $M_pS_q-Ag_2S$, $M_mX_n-Ag_2S$ (where X = Cl, Br, I, CNS;
   S = S, Se, Te; M = Tl, Cu, Pb, Hg, Cd) and a number of combinations of
   other salts;

4) membranes of amorphous semiconductors, e.g., chalcogenides of
   $Se_{60}Ge_{28}Sb_{12}$ type with the addition of Fe, Co, Ni, etc.;

5) membranes that are thin dielectric films (up to 1000 Å), e.g., SiO_2
   and Si_3N_4.

Thus, from the consideration of a list of solid membrane materials it follows that they can be single crystals, polycrystalline and amorphous and can
represent different classes of solids: solid electrolytes, semiconductors,
dielectrics. The action of an ion-selective electrode will depend, naturally,
on all these peculiarities of membrane material.

1.3. Bulk properties and electrochemical investigation of solid-state
     ion-selective electrodes

Let us consider electrochemical problems that result from the investigation of solid ion-selective electrodes (Fig.3). Obviously, the solution of an
electrolyte and a solid ion-selective electrode put in it are a three-phase
system that consists of two different liquid phases and a solid membrane in
the case of ordinary electrode, but in the case of all-solid-state electrode
it consists of a solution, a solid membrane and a solid metallic contact.
The behaviour of the whole three-phase system will be determined by both the
bulk properties of all three phases and properties of interfaces. Therefore,
to describe the action of an ion-selective electrode and to understand the

mechanism of its action investigations in the field of physics and chemistry of electrolyte solutions, physics and chemistry of solids and chemistry of surfaces are necessary.

Among the bulk properties that are to be investigated in the study of the electrochemical behaviour of ISEs one can name such as activity coefficients of single ions in solutions and structure and conductivity of a solid membrane. Investigations of phenomena at the solution-solid membrane interface the problem of a double electric layer both in the liquid phase and in the solid membrane; in this case the properties of the double electric layer in the membrane should be considered at the equilibrium of crystals both with the vacuum and with the solution. The necessity of the study of bulk properties and interface properties is seen from the consideration of a number of thermodynamic ratios. If a general case of solid ion-selective electrode with internal solution is considered, electrochemical cell with it can be written as follows:

| reference electrode | internal solution (1) | membrane | test solution (2) | reference electrode | (1) |

and in the case when a mixture of salts with common cation M_1X and M_2X is present on the solution

ϕ'_{MX}	ϕ_j	ϕ'	ϕ''	ϕ_j	ϕ''_{MX}	
M	MX	$M_1X + M_2X$		$M_1X + M_2X$	MX	M	
reference electrode	internal solution (1)	membrane	test solution	reference electrode		(2)

Then taking into account the diffusion potential ϕ_{dif} the emf E of such a cell will be expressed by the following equation

$$E = \frac{RT}{zF} \ln \frac{a''_{M_1X} + \frac{u_{M_2}}{u_{M_1}} K_{M_1 M_2} a''_{M_2X}}{a'_{M_1X} + \frac{u_{M_2}}{u_{M_1}} K_{M_1 M_2} a'_{M_2X}} \tag{3}$$

where a are activities of salts M_1X or M_2X in internal (') and test ('') solutions, K is exchange constant, u - mobilities of M_1 and M_2 cations in solid, R - gas constant, T - temperature, F - Faraday number, z - ion charge. The equation for E of the cell without taking into account the diffusion potential was made by Nicolsky first /19,20/. When $u_{M_2} \to 0$, $a_{M_2} = 0$, $K \to 0$ the equation (3) turns into a known Nernst equation

$$E = E_o + \frac{RT}{zF} \ln a_{M_1}''$$ (4)

where $a_{M_1}'' = \gamma_{M_1} c_{M_1}''$ and γ_{M_1}, c_{M_1}'' are activity coefficient and the concentration of certain ions, respectively. Thus, eqn.3 and eqn.4 show the role of bulk properties of liquid and solid states in the formation of the emf of the cell.

A great deal of data have been collected recently in connection with a number of problems mentioned above. Thus, there is a great number of papers on activity coefficients of single ions /21-24/, on the solution-solid state interface /25-28/, on physics and chemistry of solid electrolytes /29--32/. However, papers on the chemistry of solids, being a large separate field, are still little used in the study of ion-selective electrodes.

This paper shows the results of the electrochemical study of some solid ion-selective electrodes, this study being carried out on the principle of a complex investigation of physical and chemical properties of a solid membrane and electrode properties of an ion-selective electrode. Five different membranes based on different solid materials are taken as an example: 1) pressed pellets of polycrystalline salt Ag_2S, 2) pressed pellets of mechanical mixture of the $AgX-Ag_2S$ (X = Cl, Br, I) type, 3) single crystal $LaF_3:CaF_2$, 4) amorphous chalcogenide membranes, 5) thin dielectric films SiO_2 and Si_3N_4.

2. PROBLEMS OF ELECTROCHEMISTRY AND SOLID STATE CHEMISTRY IN THE INVESTIGATION OF SOLID MEMBRANES

For the understanding of the action of an ion-selective electrode one must know what particles are charge carriers, what the mechanism of charge transfer through the membrane is and in what way the charge transfer through the interface is brought about in the case of different charge carriers in different phases (solution-solid membrane or solid membrane-metallic contact). It is evident that the bulk structure of a solid membrane, structure of surface layer at the equilibrium of a membrane with vacuum, total resistance and conductivity of a membrane, ionic and electronic conductivity ratio are important characteristics of a solid. Some important properties of

solids which it isnecessaryto take into consideration for electrochemical
investigations of solid ISE are reported below.

2.1. Defects in solids

The real bulk structure of any solid material is characterized by the presence of defects that determine physical properties of a certain material,
its conductivity, in particular. Defects may be point (ionic), electronic
and macrodefects. Ionic crystals of KCl type are characterized by Schottky
point defects, crystals of AgBr type are characterized by Frenkel point defects (cation in the interstitial Ag_i^+ plus cation vacancy Ag_v^-)(Fig.4).
For example doping of the crystal by a multicharged impurity (e.g., Cd^{+2} in
AgBr) increases the concentration of cation vacancies sharply, and the doping
by an impurity with a lesser charge (e.g., Eu^{+2} in LaF_3) increases the concentration of anion vacancies.

Electron defects and, therefore, the possibility of electrons' taking
part in the charge transfer depend on the electronic structure of atoms, and
the energy of their formation can be described by the scheme (Fig.5), where
one can see the difference in the behaviour of certain kinds of solid materials. Macrodefects are imperfections of crystals due to the appearance of
associations of defects, dislocations, pores, etc., and they influence on the
total conductivity. Thus, conductivity in solids depends on the nature of a
solid and may vary over a wide range.

2.2. Double layer in solids

Structure and properties of the surfaces of solids at the equilibrium
with vacuum differ but they depend on the bulk structure and the properties
of a solid. According to a number of papers /33-35/, there is a double layer
on the surface of any ionic crystal at its equilibrium with vacuum. Similar
to double electric layer at the solution-crystal interface, different from
it in nature however. It is due to different energy of the formation of defects of opposite signs in the crystal. Thus, in the case of AgCl at the total energy of the formation of Frenkel pair 1,44 eV energies of the formation of silver vacancies Ag_v^- and interstitial silver Ag_i^+ are 0,58 eV and 0,86
eV, respectively /36/. As a result it appears that the ratio of defects of
opposite signs is violated in the surface layer and the surface turns out to
be charged in regard to the bulk. Estimation of this charge for NaCl gives
0,28 V /37/ and for AgCl - 0,1 V /36/, moreover, in the case of AgCl the sur-

face charge is negative and the depth of the location of the charge is several thousandth of a micron (Fig.6).

2.3. Solution-membrane interface

Double layer in the solid phase plays an important part in the electrochemistry of the solution-solid membrane interface and it is necessary to take its existence into account while considering the charge density, electric field and potential at the interface. This approach was successfully applied in papers /38,39/ at the study of photopotentials in silver halogenides.

3. INVESTIGATION OF PHYSICAL AND CHEMICAL PROPERTIES OF SOME TYPES OF SOLID MEMBRANES

3.1. Membranes of the $AgX-Ag_2S$ (X = Cl, Br, I) type

3.1.1. Properties of polycrystalline Ag_2S

Electrochemical properties of β-Ag_2S are of special importance in connection with the use of this compound in a large group of solid-state ion-selective electrodes, sensitive to different ions.

Silver sulphide has four modifications in different temperature ranges and pressures: monoclinic (up to $177^{\circ}C$) β-Ag_2S, BCC α-Ag_2S (from 177 to $580 \pm 20^{\circ}C$) and FCC γ-Ag_2S (above $600^{\circ}C$) as well as tetragonal δ-Ag_2S, existing at high pressure (> 10 Kbar).

Deviation from stoichiometry in Ag_2S were first found from the dependence of conductivity on the partial pressure of sulfur in the atmosphere /40/.

Ag_2S is a typical representative of non-stoichiometric crystals and its formula may be written as $Ag_{2+\delta}S$, Ag_2S being regarded as n-type semiconductor. Deviation from stoichiometry as well as a number of structural factors (micro- and macrodefects), depending on the way of preparation, could be the reason of that the data on the determination of conductivity, transfer numbers, etc., obtained by different authors differ from each other considerably. Most investigations of electrochemical characteristics were carried out at high temperature, and investigation of properties at room temperature is of a particular interest from the point of view of the study of solid ion-selective electrodes. Such research was carried out with β-$Ag_{2+\delta}S$ stoichiometric polycrystals /41/ and in equilibrium with silver ($0 \leqslant \delta \leqslant 10^{-5}$). In this paper a clear dependence of density and conductivity of samples on pressure (Fig.7) was established, and the conditions allowing to obtain samples with reproducible properties were chosen (12,5 t/sm^2). Cells I, II and III were used to obtain total (σ_{tot}), electronic (σ_{el}) and ionic (σ_{ion}) conductivity of β-Ag$_{2+\delta}$ S samples, respectively.

$$C \mid \beta- Ag_{2+\delta} S \mid C \qquad\qquad I$$

$$- Ag \mid \beta- Ag_{2+\delta} S \mid C + \qquad\qquad II$$

$$+ Ag \mid RbAg_4I_5 \mid \beta- Ag_{2+\delta} S \mid RbAg_4I_5 \mid Ag - \qquad III$$

The experimental data obtained (Table 1, Fig.8) have lead to interesting conclusions: σ_{ion} has been found not to depend on δ value, i.e. it does not depend on the deviation from stoichiometry, while σ_{tot} increases owing to σ_{el} at the increase of silver content.

Table 1. Total, ionic and electronic conductivities in ohm^{-1}cm^{-1} amd transfer numbers in β-Ag$_{2+\delta}$ S at 20°C /41/.

β- Ag$_{2+\delta}$ S	σ_{tot}	σ_{ion}	σ_{el}	t_{ion}
STOICHIOMETRIC ($\delta = 0$)	4,82 10^{-5}	3,5 10^{-5}	1.3 10^{-5}	0,73
IN EQUILIBRIUM WITH Ag ($\delta = 10^{-5}$)	5,45 10^{-4}	3,5 10^{-5}	5,1 10^{-4}	0,06

Thus, the above mentioned data show that total conductivity β-Ag$_2$S depends on macrodefects of the structure, that can be affected by the pressing conditions, and on the deviation from stoichiometry, connected with the quantity of the excess of interstitial silver Ag$_i$, while ionic conductivity β-Ag$_{2+\delta}$ S has been found not to depend on the sample composition.

3.1.2. Properties of polycrystalline AgX

AgCl, AgBr and AgI crystals are typical crystals with Frenkel type defects, they have been studied in detail and the review of papers on the properties of these salts and solid electrolytes based on them can be found in /42,43/. At ordinary pressure AgCl and AgBr have the structure of NaCl type, that does not change up to the melting point; at ordinary pressure and at room temperature AgI is a mixture of γ- and β-phases with FCC (high conductivity) and hexagonal (low conductivity) structures, respectively . This condition causes certain difficulties in the study of membranes with stable properties. The way of obtaining polycrystalline AgI precipitate, containing

155

mostly γ-AgI, from a solution is described in /44,45/. Table 2 shows general electric characteristics of polycrystalline AgX (X=Cl, Br,I) samples /46/

Table 2. Electrical properties of polycrystalline AgX (X = Cl, Br, I) samples at 25°C (inert atmosphere) /46/

Sample	$\sigma_{tot} \, 10^6$	$\sigma_{el} \, 10^8$	$\sigma_{ion} \, 10^6$	t_{ion}
AgCl	1.83	2,0	1.8	0,98
AgBr	5,82	2,3	5,8	0,99
β-AgI	1.63	7,0	1.6	0,95
γ-AgI	250	25,0	250	1.00

3.1.3. Properties of polycrystalline pressed AgX-Ag$_2$S membranes

According to the above mentioned data polycrystalline membranes based on AgX-Ag$_2$S (X = Cl, Br, I) and widely used in ion-selective electrodes are the mixtures of two different salts - solid electrolytes AgX, characterized by the presence of Frenkel type defects and n-type semiconductor Ag$_2$S the imperfection of structure of which results in the deviation from stoichiometry. Additional foundation of mobile interstitial ions occurs in AgX crystals in the case of Ag$_2$S microconcentrations ($\sim$10 ppm) according to the equation:

$$Ag_2S \rightleftharpoons S_x^{2-} + Ag_i^+ + Ag_{Ag}^+$$

However, this process resulting in the increase of the conductivity of a sample can be found only on single crystals and at small concentrations of an impurity. In ion-selective electrode membranes which are mechanical mixtures of the polycrystals of the two salts AgX and Ag$_2$S, taken in commensurable quantities the process of charge transfer and conductivity mechanism must be different. As it is shown in the papers /46,47/, conductivity of a binary mixture AgX-Ag$_2$S is not simply the sum of the conductivities of pure salts, but depends in a complicated way on the composition (Fig.9,10). The found dependences of the conductivity on the composition of AgX-Ag$_2$S membranes allowed to make a conclusion that the best mixture for ion-selective electrode membranes is the mixture of the composition 50:50 mol.%, since it containes a sufficient quantity of an active sensitive material AgX, and the ionic part of the conductivity is considerably higher than in the pure salt AgX.

3.2. Single crystal LaF$_3$:Ca^{+2} membranes

Among solid-state ion-selective electrodes the electrode selective to fluoride ions with sensitive membrane of lanthanum fluoride single crystal acti-

vated by europium is of special importance. This electrode was first suggested in paper /48/ and later different modifications of fluoride-selective
electrode appeared, e.g., with membranes of polycrystalline lanthanum fluoride in a matrix of silicone rubber /49/, of ceramics $LaF_3-EuF_2-CaF_2$ /50/
and based on bismuth fluoride /51/. Electrodes with membranes of lanthanum
fluoride single crystal with different concentrations of europium were investigated /52/, and electrodes with solid-state internal contact were suggested /53/. Lahtanum fluoride is an example of a stoichiometric crystal, in
which ions of a basic element and an impurity are statistically distributed
on structurally equivalent positions. Doping by Eu^{2+} ion having a lesser positive charge than a basic La^{3+} ion results in the occurance of defects in
the anionic sublattice or so-called anti-Frenkel type defects, which causes
the increase of conductivity in LaF_3 single crystal. Usually membrane LaF_3
doped with 0,8 at.% Eu^{+2} is used in lanthanum fluoride electrodes. However
other double-charged ions can play the same part in the LaF_3 lattice. That
is why an electrode with membrane of $LaF_3:Ca^{+2}$ was suggested in paper /54/.
Influence of impurities on the crystal conductivity is seen from table 3.

Table 3. Conductivity of pure and doped LaF_3 single crystals at $25^{\circ}C$.

LaF_3 /55/	$LaF_3:Eu^{+2}$ (o,8 %) /54/	$LaF_3:Ca^{+2}$ /54/		
		1,2 %	2,4 %	4,7 %
σ $ohm^{-1}_{cm^{-1}}$ $4 \cdot 10^{-7}$	$9 \cdot 10^{-6}$	$3 \cdot 10^{-6}$	$7,5 \cdot 10^{-6}$	$1.3 \cdot 10^{-5}$

Activation energy of ionic conductivity for all the membranes calculated
from the temperature dependence found (Fig.12) turned out to equal $0,35 \pm$
$\pm 0,02$ eV. In terms of the found value one can make an assumption about an
identical character of ionic conductivity in the crystals under consideration which evidently depends little on the nature of an impurity. Earlier it
was shown /56/ that in fact the total conductivity of LaF_3 crystals consists
fully of ionic conductivity. Electrodes with the membrane of $LaF_3:Ca^{+2}$ did
not yield to common lanthanum fluoride electrodes in their electrode properties, however, preparation of such membranes is an easy task.

3.3. Semiconducting chalcogenide membranes

Chalcogenide glasses doped with metal impurities are progressive membrane materials for obtaining all-solid-state ion-selective electrodes. Cop-
per- and lead-containing vitreous alloys based on arsenic chalcogenides we-

re found to be sensitive to Cu^{+2} and Pb^{+2} ions, respectively /57/,
$Fe_n Ge_{28} Sb_{12} Se_{60}$ chalcogenide composition (where n = 1.3-2,0) were used as
ISE membranes for the determination of Fe^{+3} or Cu^{+2} ions in chloride, nitrate and perchlorate media /58,59/.

At present there is no common point of view on the mechanism of action of
the above mentioned electrodes. Both adsorption /57,59/ and redox /58/ mechanisms of potential formation were suggested. To solve this problem a complex investigation of the properties of vitreous alloys /60,61,62/ (where
x = 0-2,2) was undertook and it involved dc conductivity measurements, measurements of the edge of optical absorption and magnetic susceptibility,
the study of local environment and valence state of iron atoms by Mössbauer
spectroscopy on ^{57}Fe along with potentiometric measurements of electrode behaviour of the membranes of solid-state ISEs based on the above mentioned
material in iron (III) chloride, nitrate, copper(II) and iron(II) chlorides,
lead and cadmium nitrate solutions , in redox media.

The obtained ISEs with vitreous $Fe_{2,0}(Ge_{28}Sb_{12}Se_{60})_{98,0}$ membranes were
found to be sensitive to the activity change of Fe^{+3} and Cu^{+2} ions in chloride and nitrate media.

Fig.13 shows the electrode function of such electrodes in $FeCl_3$ solutions (constant ionic strength was created by 1-M KCl, pH=1,6). Slight differences in standard potentials (no more than 5-6 mV) show great stability in
action and reproducibility of electrode potentials. In $2 \cdot 10^{-4}$-$3 \cdot 10^{-2}$ mole/l
concentration region the slope is 56-60 mV/decade, i.e., close to theoreti-
cal Nernst slope for the monocharged electrode function. At the further increase of the solution concentration decrease of the slope to 40-50 mV/deca-
de occurs and in more dilute solutions (10^{-5} M and less) electrode potential
does not depend on the concentration.

Electrode behaviour of ISE obtained in redox media (solutions $FeCl_2/FeCl_3$,
1 M KCl, pH=1,6) is shown in Fig.14, where one can see that the potentials
of the investigated electrodes are considerably different from equilibrium
redox potentials of platinum electrode. In chloride solutions with constant
ion ratio Fe^{+2}/Fe^{+3} platinum electrode potential remains constant, while
the potential of the investigated ISEs increases as the iron(III) concentration in solution rises (Fig.15).

Ion-exchange process at the iron(III) salt solution/iron-containing chalcogenide glass interface cannot play a determining part in the potential

formation mechanism, since the slope of potential function of an ion-exchange electrode in this case must be of the order 20 mV/decade in accordance with the equation:

$$E = E^o_{ox} + \frac{RT}{3F} \ln \frac{a_{Fe^{+3}_L}}{a_{Fe^{+3}_S}} \qquad (5)$$

The data from papers /58,59,62/ give evidence that the slope of electrode function of the investigated solid-state ISE is close to theoretical Nernst slope for monocharged electrode function (59 mV/decade).

In the redox mechanism of potential formation suggested in paper /58/ it was considered that the activity of the reduced form $a_{Fe^{+2}}$ near the surface of the electrode can be maintained at the same level, independent of the activity of Fe^{+3} ions in solution, so:

$$E = E^o_{redox} + \frac{RT}{F} \ln \frac{a_{Fe^{+3}}}{a_{Fe^{+2}}} \qquad (6)$$

In consequence the investigated electrodes show the dependence on $a_{Fe^{+3}}$ with the slope of the order 59 mV/decade. From such a mechanism one can suppose that the investigated ISEs behave as redox electrodes of noble metals. Data, shown in Fig.14 and 15 /62/ contradict with the point of view described above.

Data on dc conductivity obtained in /60-62/(nonlinearity in temperature dependence),on the shift to lesser energies in the optical spectra(Fig.16,17), on ^{57}Fe Mössbauer spectra and on magnetic susceptibility, showed that in $Fe_x(Ge_{28}Sb_{12}Se_{60})_{100-x}$ ($x \leq 2,2$) iron is mostly in Fe^{+2} state and electronic properties of certain vitreous alloys are determined by deep $Fe^{+2}(d^6)$ impurity donor centres. Band model of iron-containing chalcogenide glasses is shown in Fig.18. In accordance with the results of these solid-state investigations one can assume that the potential determining process is the electron exchange between donor centres $Fe^{+2}(d^6)$ in the surface layer of vitreous membrane and electron acceptors in solution (Fe^{+3} or Cu^{+2} ions):

$$Fe^{+3}_L + Fe^{+2}(d^6) \rightleftharpoons Fe^{+2}_L + Fe^{+3}(d^5) \qquad (7)$$

At certain assumptions electrode potential in this case may be expressed by the equation:

$$E = E^o + \frac{RT}{F} \ln a_{Fe^{+3}_L} \qquad (8)$$

3.4. Thin dielectric SiO_2 and Si_3N_4 films as membranes in ion-selective field effect transistors (ISFET)

Thin SiO_2 and Si_3N_4 films (1000 Å) are used as membranes in ion-selective field effect transistors (ISFETs) sensitive to the change of pH of a solution. The scheme of such an electrode and the simplest way of the distribution of a potential are shown in Fig.19 /63/. Dielectric SiO_2 and Si_3N_4 films differ from all the membrane materials considered above by exceptionally high resistance (10^{13} ohm). Electrochemical investigation of the behaviour of ISFET involves, in particular, the study of EDS-structures (electrolyte - - dielectric - semiconductor). There is a great number of papers on the study of such structures (see, e.g. /64/) in the literature on physics. Investigations dealing exactly with ISFET are described in a number of review papers and articles (see, e.g. /63,65-67/). Ion-sensitivity mechanism of ISFET cannot be considered to be established at present. Here, as well as in the investigations of ordinary ion-selective electrodes, a complex of phenomena must be studied, involving bulk properties of membrane material and their dependence on the membrane defect structure. In this connection a number of investigations /68-70/ on the determination of SiO_2 and Si_3N_4 pH sensitivity mechanism was undertaken. The fact of the dependence of hydrogen concentration in Si_3N_4 on the conditions of film treatment (Fig.20) /71/ was used. It appeared that pH sensitivity changes with the change of hydrogen content in Si_3N_4 films.

4. SUMMARY

To understand the processes taking place in the action of solid ion-selective electrodes a complex study of a number of electrochemical problems is necessary, connected both with the surface and bulk phenomena. One of such important problems is the electrochemical study of bulk properties of solid materials used as ion-selective electrode membranes. The bulk defect structure is shown to play an important role in electrochemical behaviour of the ion-selective electrodes.

REFERENCES

1. R.P.Buck, V.R.Shepard, Jr., Anal.Chem. 46,2097(1974)
2. Ion selective electrodes. Ed.by R.A.Durst, National Bureau of Standarts, Spec.Publ. 314, 1969.

3. G.J.Moody, J.D.R.Thomas, Selective Ion Sensitive Electrodes. Merrow Technical Library, Wotford, England, 1971

4. N.Lakshminarayanaiah, Membrane Electrodes, Academic Press, New York, 1976

5. B.P.Nicolsky, E.A.Materova, Ion-Selective Electrodes, Khimia, Leningrad, 1980

6. Ion-selective electrodes in analytical chemistry. vol.1, Ed.by H.Freiser, Plenum Press, New York, 1978

7. J.Koryta, Ion-selective Electrodes. Cambridge University Press, Cambridge, England, 1975

8. D.Midgley, K.Torrance, Potentiometric Water Analysis. John Wiley & Sons, New York, 1978

9. Ion-selective Electrode Methodology. Ed.by A.K.Covington, CRC Press, Inc., 1979

10. Ion-selective electrodes. Ed.by E.Pungor, Akadémiai Kiadó, Budapest, 1977

11. Ion-selective electrodes. Ed.by E.Pungor, Akadémiai Kiadó, Budapest, 1978

12. Ionnyi obmen i ionometria. Ed.by B.P.Nicolsky, v.1, Leningrad Univ.Press, Leningrad, 1976

13. Ionnyi obmen i ionometria. Ed.by B.P.Nicolsky, v.2, Leningrad Univ.Press, Leningrad, 1979

14. R.P.Buck, Anal.Chem. 48,23R(1976)

15. R.P.Buck, Anal.Chem. 50,17R(1978)

16. B.P.Nicolsky, E.A.Materova, A.L.Grecovich, Zh.anal.khim. 30,2223(1975)

17. B.P.Nicolsky, E.A.Materova, A.L.Grecovich, Elektrokhimia 13,740(1977)

18. Pure and Appl.Chem. 48,127(1976)

19. B.P.Nicolsky, Zh.fiz.khim. 10,495(1937)

20. B.P.Nicolsky, Zh,fiz,khim. 27,1727(1953)

21. R.G.Bates, R.A.Robinson, in Ion-selective electrodes (E.Pungor ed.). Akadémiai Kiadó, Budapest, 1978

22. R.G.Bates, R.A.Robinson, Pure Appl.Chem. 37,575(1974)

23. B.Fehrmann, M.Moritz, M.Breitenbach, Chim.Anal.(Warsaw)24,775(1979)

24. V.A.Rabinovich, A.E.Nikerov, V.P.Rotshtein, Teor.i eksp.khim. 8,32(1972)

25. E.P.Honig, G.H.Th.Hengst, J.Coll.Interface Sci. 31,545(1969)

26. T.B.Grimley, N.F.Mott, Disc.Faraday Soc., 3,1(1947)

27. J.Lyklema, J.Overbeek, J.Coll.Sci. 16,585(1961)

28. P.L.Levine, S.Levine, A.L.Smith, J.Coll.Interface Sci. 34,549(1970)

29. T.Takahashi, J.Appl.Electrochem. 3,79(1973)

30. Physics of Electrolytes. Ed.by S.Hladik, Acad.Press, London-N.Y., 1972

31. A.N.Myrin, Khimia nesovershennykh ionnykh kristallov. Leningrad Univ. Press, Leningrad, 1975

B.B.Owens, J.Electrochem.Soc. 117,1536(1970)

33. K.Lehovec, J.Chem.Phys. 21,1123(1953)

34. J.D.Eshelby, Newey C.W.A., P.L.Pratt, A.B.Lidiard, Phil.Mag. 3,75(1958)

35. I.M.Lifshitz, Ya.E.Gegysin, Fisika tv.tela 7,62(1965)

36. L.Slifkin, W.Mc.Gowan, A.Fukai, J.-S.Kim, Photographic Science and Engineering, 11,79(1967)

37. V.N.Chebotin, M.V.Perfilev, Elektrikhimia tverdykh elektrolitov. Khimia, Moskva, p.149, 1978

38. R.K.Rhodes, R.P.Buck, J.Electroanal.Chem. 103,19(1979)

39. R.K.Rhodes, R.P.Buck, J.Electroanal.Chem. 103,29(1979)

40. C.Tubandt, H.Reinhold, Z.Elektrochem. 37,589(1931)

41. Yu.G.Vlasov, Yu.E.Ermolenko, Hung.Sci.Instr. in press (1980)

42. K.Funke,Progress in Solid State Chem. 11,345(1976)

43. Progress in Surface and Membrane Science. v.6, Ed.by J.F.Damelli, M.D.Rosenberg, D.A.Cadenhead, Academic Press, N.Y.-London, p,1-51, 1973

44. G.Burley, Amer.Mineral. 48,1266(1963)

45. E.J.W.Verwey, H.R.Krugt, Z.phys.Chem. A167,142(1933)

46. Yu.G.Vlasov, S.B.Kocheregin, in Ion-selective Electrodes (E.Pungor ed.). Akadémiai Kiado, Budapest, 1978

47. Yu.G.Vlasov, S.B.Kocheregin, in Ionnyi obmen i ionometria (B.P.Nicolsky ed.). Leningrad Univ.Press, Leningrad, 1979

48. M.S.Frant, J.W.Ross, Science 154, 1553(1966)

49. A.M.G.Macdonald, K.Toth,Anal.Chim.Acta 27,63(1968)

50. H.Hirata, M.Ayuswa,Chem.Phys.Lett. 31,1451(1974)

51. M.S.Frant, Patent USA 3.431.182 (1968)

52. R.R.Taracyantz, R.N.Potzepkina, V.P.Rose, E.A.Bondarevskaya, Zh.anal. khim. 27,808(1972)

53. O.O.Lyalin, M.S.Tyraeva, Zh.anal.khim. 31,1879(1976)

54. Yu.G.Vlasov, Yu.E.Ermolenko, V.V.Kolodnikov, M.S.Miloshova, Zh.anal.khim. 35,691(1980)

55. A.Sher, R.Solomon, K.Lee, M.W.Muller, Phys.Rev. 144,593(1966)

56. L.E.Nagel, M.O'Keeffe, in Fast Ion Transport in Solids (van Gool W. ed.). North-Holland Publ.Co., Amsterdam-London, 1973

57. A.E.Owen, J.Non-cryst.Solids 35&36,999(1980)

58. C.T.Baker, I.Trachtenberg, J.Electrochem.Soc. 118,571(1971)

59. R.Jasinski, I.Trachtenberg, J.Electrochem.Soc. 120,1169(1973)

60. E.A.Bychkov, Yu.G.Vlasov, Z.U.Borisova, Fiz.i khim.stekla, 4, 335(1978)

61. A.A.Andreev, Z.U.Borisova, E.A.Bychkov, Yu.G.Vlasov, J.Non-cryst.Solids
    35&36,901(1980)

62. Yu.G.Vlasov, E.A.Bychkov, Hung.Sci.Instr. in press (1980)

63. Yu.G.Vlasov, Zh.prikl.khim. 52,3(1979)

64. Voprosy elektroniki tverdogo tela. Ed.by L.P.Strachov, Leningrad Univ.
    Press, Leningrad, 1978

65. J.Janata, R.J.Huber, Ion-selective Electrode Rev. 1,31(1979)

66. W.M.Siu, R.S.C.Cobbold, IEEE Transact.on electron devices ED-26,1805(1979)

67. R.P.Buck, D.E.Hackleman, Anal.Chem. 49,2315(1977)

68. Yu.G.Vlasov, Yu.A.Tarantov, A.P.Baraban, V.P.Letavin, Zh.prikl.khim.
    53,1980(1980)

69. Yu.G.Vlasov, Yu.A.Tarantov, A.P.Baraban, V.P.Letavin, Zh.prikl.khim.
    53,2345(1980)

70. Yu.G.Vlasov, A.V.Bratov, V.P.Letavin, Proceedings of the Third Scietific
    Session on Ion-Selective Electrodes, Mátrafüred, Hungary, 1980

71. V.I.Belyi, F.A.Kuznetsov, T.P.Smirnova, L.V.Chromova, L.Kh.Kravchenko,
    Thin Solid Films 37,L39(1976)

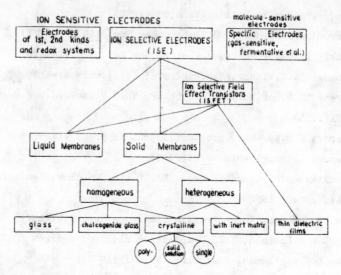

Fig.1. Classification of electrodes for determination of concentration of
different spiecies.

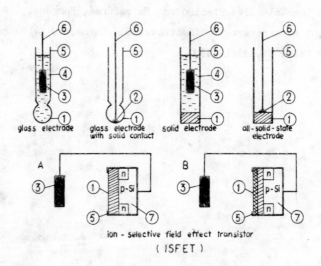

ion - selective field effect transistor
(ISFET)

Fig.2. Some types of solid ion-selective electrodes

1 - membrane, 2 - solid contact, 3 - reference electrode, 4 - internal
solution, 5 - glass or plastic body, 6 - wire, 7 - p-Si; A - ISFET
with thin film dielectric membrane (SiO_2, Si_3N_4), B - ISFET with solid
membrane on the thin dielectric film (SiO_2, Si_3N_4).

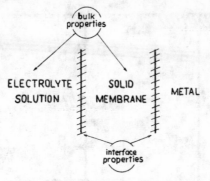

Fig.3. Solid ion-selective electrode as three phase system.

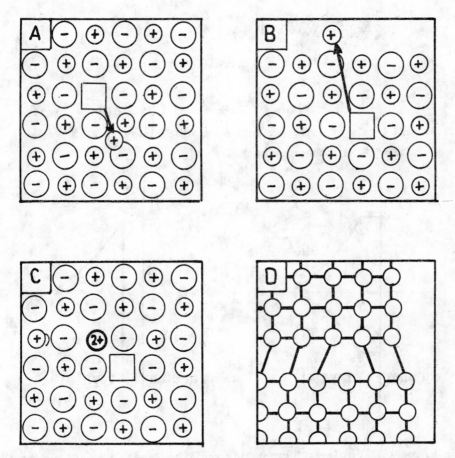

Fig.4. Defects in solids. A - Frenkel type defect, B - Schottky type defect,
C - multicharge impurity ion, D - dislocation.

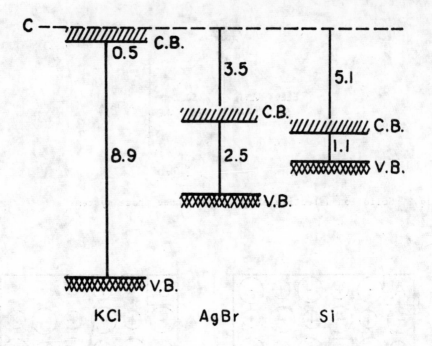

Fig.5. Band models of different solids (R.P.Buck in /9/). V.B. - valence band, C.B. - conduction band.

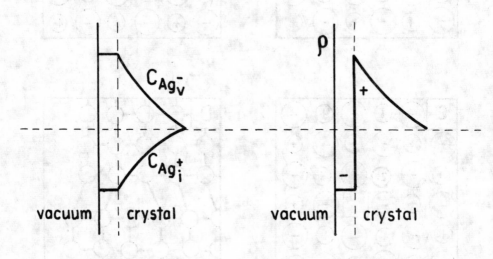

Fig.6. Double layer in AgCl crystal in equilibrium with vacuum. A - concentration profiles, B - charge density (ρ) profile, $C_{Ag_i^+}$ and $C_{Ag_v^-}$ - concentrations of interstitial ions and vacancy, respectively.

166

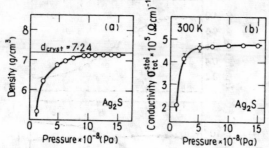

Fig.7. Density (a) and total conductivity (b) dependence of stoichiometric -Ag_2S polycrystalline samples on pressure /41/.

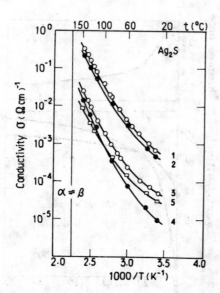

Fig.8. Temperature dependence of β-$Ag_{2+\delta}S$.conductivity.
Total and electronic conductivities at $\delta = 10^{-5}$ (1,2) and $\delta = 0$ (3,4), and ionic conductivity (5) at $0 \leqslant \delta \leqslant 10^{-5}$.

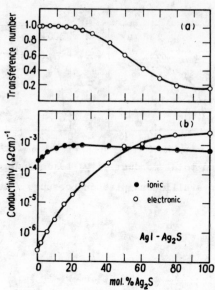

Fig.9. Transfer number (a) and electronic and ionic conductivities (b) of AgI-Ag$_2$S at 25°C.

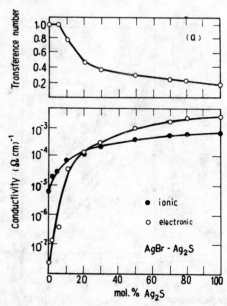

Fig.10. Transfer number (a) and electronic and ionic conductivities (b) of AgBr-Ag$_2$S at 25°C.

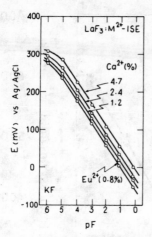

Fig.11. Response of doped LaF$_3$ single crystal electrodes /54/.

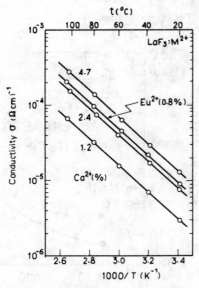

Fig.12. Conductivity dependence of doped LaF$_3$ single crystals on temperature /54/.

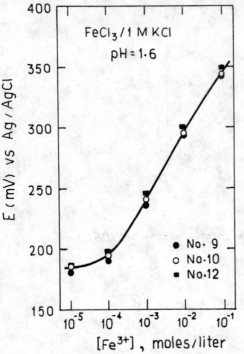

Fig.13. Response of ion-selective
electrodes (no.9,10,12) with
$Fe_2(Ge_{28}Sb_{12}Se_{60})_{98}$ vitreous
membranes in $FeCl_3$ solutions
/62/.

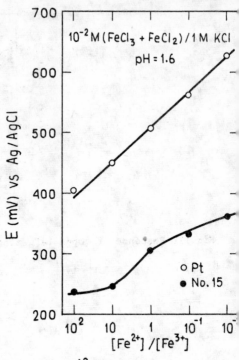

Fig.14. Fe^{+3}-electrode behaviour in
$FeCl_2/FeCl_3$ solutions /62/.

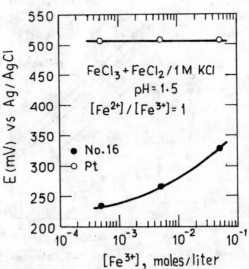

Fig.15. Fe^{+3}-electrode behaviour in the media with constant $FeCl_2/FeCl_3$
ratio, but with different Fe^{+3} ions concentration.

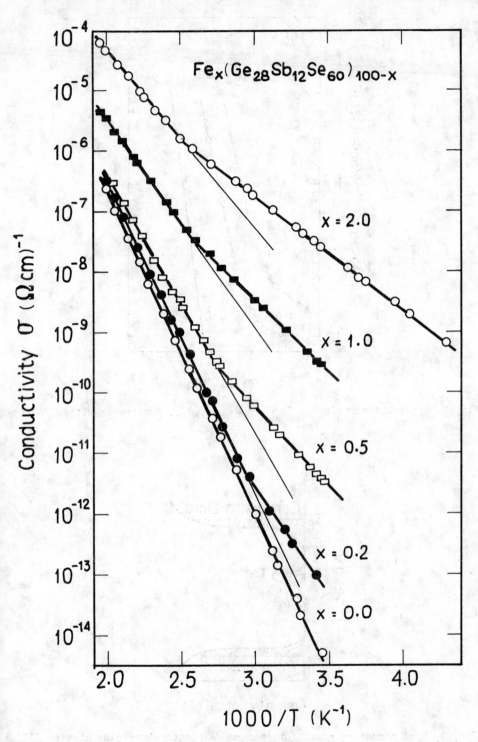

Fig.16. D.c. conductivity for $Fe_x(Ge_{28}Sb_{12}Se_{60})_{100-x}$ vitreous alloys /60/.

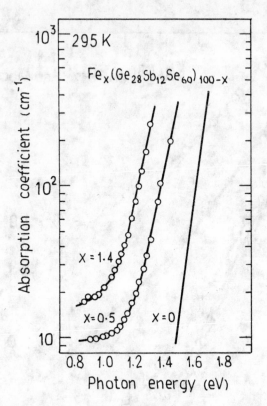

Fig.17. Optical absorption in $Fe_x(Ge_{28}Sb_{12}Se_{60})_{100-x}$ vitreous alloys /61,62/.

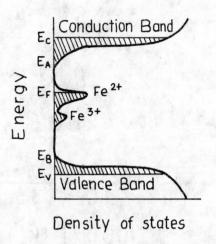

Fig.18. Band structure model for iron containing chalcogenide glasses /62/.

172

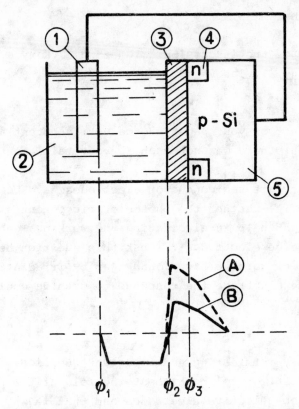

Fig.19. Schematic diagram of ion-selective field effect transistor (ISFET)
and potential profiles in solutions with different (A and B) concentration. 1 - reference electrode, 2 - test solution, 3 - thin film
dielectric membrane, 4 - n-Si, 5 - p-Si; ϕ_1, ϕ_2, ϕ_3 - interface
potentials.

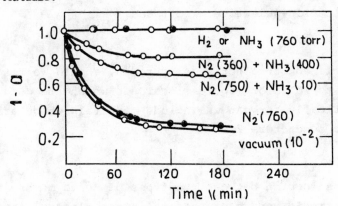

Fig.20. The influence of the ambient on the decomposition of N-H groups in
Si_3N_4 films during annealing at $1000°C$ /71/.
a - the relative degree of N-H group dissociation.

173

QUESTIONS AND COMMENTS

Participants of the discussion: Y.Umezawa, J.Koryta, J.D.R.
Thomas, G.Nagy, Y.Vlasov

Question:
In the abstract of your paper photoelectric potentials were
mentioned, but there was not much in your talk about this
effect.

In recent years semiconductor electrodes have been spreading
The question is, whether the photoelectric potentials you
mentioned might help in elucidating the working mechanism
of pH-responsive electrodes ? I ask this question because
in some earlier work we have found that by irradiating the
silver iodide electrode the response to cyanide could be
improved.

Answer:
This is a very interesting and important question. Originally,
it was my intention to speak about the role of the photo-
-effect of ion-selective electrodes and that is why it was
mentioned in the abstract. This photo-effect is particularly
pronounced with single crystal electrodes. After reading the
papers of Dr.Buck on this subject, I decided to leave this
part out of my paper. The interpretation of the photo-effect
seems to be very useful in the elucidation of the working
mechanism of ion-selective electrodes.

Question:
When we define ion-selective electrodes, should we restrict
the field to solid or liquid electrodes which are conducting
electrolytically, or can we include materials which are
electronic conductors ?

Answer:
One can distinguish between two types of solid membranes, one
being electronic conductors and the other electrolytical ones.

In the bulk of the membrane you can differentiate between two charge transfer mechanisms: you have different types of charge carriers, they can be cations or particles which carry electrons. Both kinds of matrices can work as ion-selective electrodes under appropriate conditions.

Comment:
We must put a definite boundary between ion-selective electrodes and classical electrochemical systems, both metallic and semiconducting. I think we should exclude systems based on redox processes, because otherwise we would have a diffusion layer the end of which is not known.

Answer:
If you take such a simple material as Hg_2SO_4 it shows an n-type conductivity.

Comment:
I think this problem has been discussed quite deeply by Dr.Buck. He has shown that the type of working mechanism greatly depends on the conditions. If we do not define these conditions exactly, then we run into serious problems at least from the point of view of nomenclature.

Question:
In connection with the LaF_3 electrode it was mentioned that any impurity would serve the purpose, taking impurity as a relatively small amount of a foreign material, and yet you use quite large quantities of calcium ?

Answer:
It is really not important what kind of foreign material you put into the single crystal of LaF_3 but it is important that this material should carry two positive charges and it should be dispersed inside the crystal in the vacancies, which provides a good conductivity for the single crystal. In my case, the LaF_3 single crystal could be doped with up to 4-5% calcium, and it was homogeneous and stable. At higher calcium concentrations, however, the crystal decomposes and becomes unstable. The role of calcium/II/ is the same as that of

europium/II/, but the concentration of the latter is about 0.8%. This concentration must be kept constant quite rigorously. In our case the calcium/II/ level in the crystal is not critical, and the conductivity is about equal to that of the europium/II/ doped crystal.

Question:
How was the conductivity of these membranes measured ?

Answer:
Different methods were used, according to whether the total, electronic or ionic conductance was to be measured. All these methods have already been published and I can give you the list of the papers.

DISCUSSION LECTURES

VALINOMYCIN BASED SILICONE RUBBER MEMBRANE ELECTRODES FOR CONTINUOUS MONITORING OF POTASSIUM IN URINE

D. AMMANN, P. ANKER, H.-B. JENNY AND W. SIMON

Swiss Federal Institute of Technology, Department of Organic Chemistry,
CH-8092 Zürich, Switzerland

ABSTRACT

A flow-through electrode system based on a silicone rubber membrane with valinomycin as the ion-selective component is described. In contrast to the conventional valinomycin based PVC membrane electrodes this electrode allows continuous measurement of K^+-activities in undiluted urine samples. A correlation of the K^+-concentrations determined by the ion-selective electrode with those obtained using a flame photometer yields a residual standard deviation of ±1.4 mM over the 15 to 50 mM K^+ concentration range.

INTRODUCTION

The continuous measurement of Na^+- and K^+-concentrations in undiluted urine of patients in intensive care units is of great interest [1]. Only recently Na^+-selective PVC membrane electrodes which allow the measurement of Na^+-activities in undiluted urine samples were described [2]. However, extended studies on the measurement of K^+-activities in undiluted urine samples with conventional valinomycin based PVC membranes have not been reported so far. Today, the use of such PVC membranes is restricted to measurements in diluted urine samples [3, 4]. Here we report on the use of valinomycin based silicone rubber membranes in long-time studies on bedside-monitoring of undiluted urine.

RESULTS

A comparison of K^+-measurements in undiluted urine samples obtained potentiometrically with PVC membrane electrodes and flame photometrically clearly indicated too low K^+-readings for the electrode [5]. This could not be rationalized under the assumption of interfering cations. Possible explanations of low electrode readings are interference by lipophilic sample anions and/or the presence of complexing agents for potassium in

urine. Complex formation seems unlikely since an acidification
of the urine samples to about pH = 2 did not improve the situation [5]. In 1976 valinomycin based silicone rubber membranes
were described which, in comparison to the corresponding PVC
membranes, show a drastically reduced anion interference [6].
Only recently similar silicone rubber membranes have been prepared by another technique and have been used in K^+-activity
measurements in undiluted urine [7].

In Figure 1 the anion interference of K^+-selective membranes
with silicone rubber or PVC as the membrane matrix are compared.
It is evident that the PVC membranes suffer to a larger extent
from interference by the lipophilic thiocyanate anion than the
silicone rubber membrane. The assumption of anion interference
as the origin of too low electrode readings is underlined
since this silicone rubber membrane electrode leads to good
correlations with flame photometry [7].

Only recently these optimized membranes were tested on two
patients over about three days in the intensive care unit of an
hospital in Basle (PD Dr. G. Wolff, Kantonsspital Basel). Minia-
turized membrane electrodes (see [2, 7]) were used in a flow-
through system (see Figure 2). The outflow of a bladder catheter
was fed directly into the electrode system. A peristaltic pump
was used to suck the urine continuously from a syphon (as a
barrier for sterilization) through the electrode system. Different valves enabled the insertion of calibration solutions and
the withdrawal of samples for comparison measurements. The
slope of the electrode cell assembly was calibrated every
6 hours and after every hour of continuous measurement on urine
a one-point calibration was performed. For further experimental details on calibration solutions, the calibration technique, and the computation of the concentrations see [2] and
[7]. In Figure 3 the K^+-concentration profiles obtained by continuous measurement on the urine of two patients are presented.
Comparison measurements using a flame photometer were run every
hour on samples collected over six minutes (dots in Figure 3).
The correlation of the results obtained by potentiometry and
flame photometry shows that with clinical routine work in mind
the electrode system leads to K^+-values of sufficient precision
(Figure 4). The deviation from the regression line is comparable
to the standard deviation of measurements obtained on discrete
samples [7] even though the electrode readings correspond to
instantaneous values whereas the flame photometer data represent
6-minutes averages.

DISCUSSION

The results presented here obviously open new perspectives
for the continuous monitoring of K^+ in urine. However, in order
to assess a diagnostic value of such electrode measurements in
the urine of patients further experiments are necessary. The

flow-through system as used in this contribution seems to be unsuitable for routine work because during times of small urine flow air is aspirated. By introducing the electrodes directly into the syphon of the urine collecting vessel or into the bladder catheter (see Figure 2) the continuous contact of the membrane with urine would be assured. A definite problem is the dead volume (~35 ml) between the kidney and the syphon which can delay measurement relative to excretion by a variable amount of time. Since two-thirds of this volume are due to the urinary tract this problem could only be circumvented by introduction of the electrode directly into the bladder or even the kidney. At the time of writing this is hardly possible since certain ion-selective liquid membrane electrodes can not be sufficiently sterilized. The question of a sterilizable catheter electrode must be seen in the light of the risks involved in using catheters as such. Results obtained with the described measuring system should indicate whether efforts in the direction of a bladder or kidney catheter would be justified.

ACKNOWLEDGEMENT

This work was partly supported by the Swiss National Science Foundation.

REFERENCES

1. H. Keller, R. Niederer, and W. Richter, Biomed. Techn. 25, 42 (1980).
2. H.-B. Jenny, D. Ammann, R. Dörig, B. Magyar, R. Asper, and W. Simon, Mikrochim. Acta, in press.
3. A. Pelleg and G. B. Levy, Clin. Chem. 21, 1572 (1975).
4. J. H. Ladenson, Clin. Chem. 25, 757 (1979).
5. D. Ammann, H.-B. Jenny, P. Anker, U. Oesch, and W. Simon, Proceedings of the Symposium on Theory and Application of Ion-Selective Electrodes in Physiology and Medicine II, Max-Planck-Institut für Systemphysiologie, Dortmund, GFR, July 28-30, 1980, in press.
6. E. Lindner, P. Wuhrmann, W. Simon, and E. Pungor, Proceedings of the 2nd Symposium on Ion-Selective Electrodes, Mátrafüred, Hungary, October 18-21, 1976, Ion-Selective Electrodes (E. Pungor and I. Buzás (Eds.)), Akadémiai Kiadó, Budapest, 1977, p. 159.
7. H.-B. Jenny, C. Riess, D. Ammann, B. Magyar, R. Asper, and W. Simon, Mikrochim. Acta, in press.

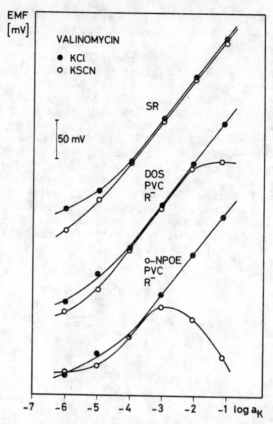

Figure 1. Response of valinomycin based K⁺-selective PVC and silicone rubber membrane electrodes in sample solutions containing anions of different lipophilicity (see also [6]). Composition of the membranes: 2.5 wt.-% valinomycin (Fluka AG, Buchs, Switzerland), 83.0 wt.-% Silopren K 1000 (Bayer AG, Leverkusen, BRD), 14.5 wt.-% cross-linking agent Siloprenvernetzer KA-1 (Bayer AG, Leverkusen, BRD) (upper curve); 0.6 wt.-% valinomycin, 0.24 wt.-% potassium tetra-(p-chloro-phenyl)borate (R⁻), 66.16 wt.-bis-(2-ethyl-hexyl)sebacate (DOS), 33 wt.-% polyvinyl chloride (PVC) (middle curve); 0.6 wt.-% valinomycin, 0.24 wt.-% potassium tetra-(p-chloro-phenyl)borate, 66.16 wt.-% o-nitro-phenyl-n-octyl ether (o-NPOE), 33 wt.-% polyvinyl chloride (lower curve).

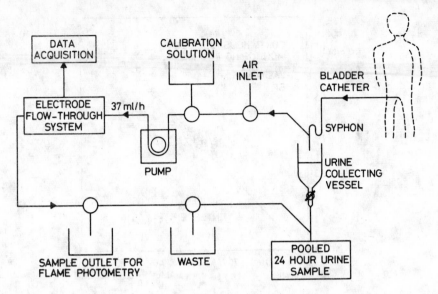

Figure 2. Scheme of extracorporal urine flow for continuous monitoring of the K^+-activity in an intensive care unit.

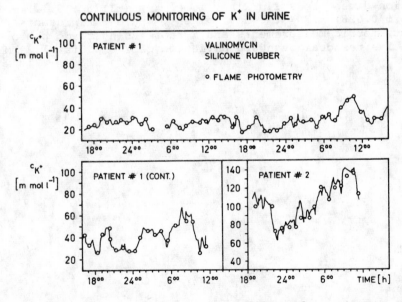

Figure 3. Potassium concentration profiles of continuous on-line measurements on the urine of two patients in an intensive care unit. The outflow of a bladder catheter is fed directly into an electrode flow-through system. Comparison measurements using a flame photometer were run every hour (dots).

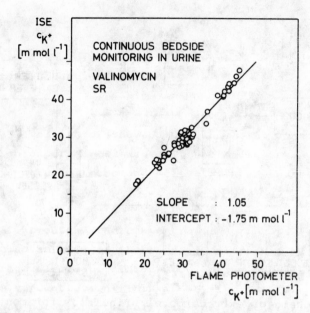

Figure 4. Correlation of results of K$^+$ measurements in urine obtained by potentiometry (ISE, during continuous monitoring, instantaneous values) and flame photometry (6-minutes averages). Membrane composition see Figure 1, upper curve. A linear regression gives the solid line, i. e. Y = (-1.75 ± 0.12) + (1.05 ± 0.026) · X (std. dev. given), where Y represents the potentiometric measurements and X those using flame photometry (mM). The residual standard deviation of Y on X is ±1.39 mM.

COMMENT

Comment:

A comparison of the standard deviations of potentiometric and flame photometric potassium determinations respectively show excellent agreement. This fact indicates that the potentiometric data can be relied upon.

184

THE SODIUM ION-SELECTIVE GLASS ELECTRODE AS A NEW MONITOR IN THE BAYER-PROCESS

I. BERTÉNYI AND L. TOMCSÁNYI

R/D Centre of Hungarian Aluminium Corporation, Budapest, Hungary

ABSTRACT

Sodium ion-selective glass electrodes can be used to monitor the sodium content of different aluminate solutions. The variation of the activity coefficient, the liquid-junction potential and the selectivity of the sodium electrode are discussed considering the concentrated Bayer liquors.

INTRODUCTION

The caustic sodium and aluminate concentrations of different aluminate liquors are the most characteristic data in the Bayer-process at present.

The widely used continuous conductometric measurements/1, 2/ are not sufficient for the proper analytical characterization of the liquors. The well-known difficulties of the use of a complicated measuring system in extremely corrosive industrial media and the requirements of high precision encouraged us to search for other methods.

The application of a sodium ion-selective glass electrode could be useful to solve this problem provided proper conditions for the monitoring system can be maintained.

EXPERIMENTAL

Radelkis OP-Na-7113, OP-Na-0744P and Polymetron 8440 sodium ion-selective glass electrodes as well as different Ag/AgCl and Hg/HgO reference electrodes were used in the cells constructed for the stationary and continuous measurements. The reference electrodes were prepared as recommended in Ives and Janz's book /3/. Radiometer pH M64 and Radelkis OP-208 precision pH meters connected to recorders were applied to determine the electromotive force. Ultrathermostate and peristaltic pump were also used for the measurements with flow-through cells. All chemicals used were of analytical grade.

RESULTS AND DISCUSSION

Four different disturbing effects must be considered in the case of ion-selective measurements in solutions containing high concentration of ions to be determined: the change of activity coefficient, the liquid-junction potential, the electrode function /selectivity and the effect of the chemically agressive media as well.

The mean activity coefficient in a sodium hydroxide solution can be determined by the measurement of the electromotive force of the glass $Na_s \| NaOH \| HgO_s | Hg_s$ cell without liquid-junction.

The details of this method can be find elsewhere e.g. /4/.

There is quite a large difference between our results and the activity coefficient data in /5/ specially at the higher concentration range. In this concentration range the electrode junction may be sub-Nernstian due to the anion effects on glass membranes according to Buck's investigation /6/. As the concentration-activity function is nearly linear in the concentration range of 7,5-10 M NaOH, the semilogarithmic calibration curve /curve 3 in Fig.1/ can be applied to evaluate the sodium concentration.

The liquid-junction potential can be calculated from the electromotive force of the cell

$$Na_s \;\|\; NaOH \;|\; NaOH \;\|\; HgO_s \;|\; Hg_s$$
$$\text{glass} \qquad mx \qquad m$$

knowing the activity coefficients. The liquid-junction potential - sodium concentration function is linear on the semilogarithmic paper, therefore the calibration curve is also linear, however, the slope is not the same /curve 4 in Fig. 1/.

In concentrated sodium aluminate solutions only the aluminate ions may disturbe the operation of the sodium ion-selective electrode. According to the experiments neither the activity coefficient nor the liquid-junction potential are influenced by the aluminate ions.

The sodium content of different Bayer plant liquors were determined by sodium ion-selective electrode measuring the electromotive force of the next cell:

$$\frac{Na}{glass} \;\bigg\|\; \text{plant liq.} \;\bigg|\; \frac{\text{standard}}{\text{liq.}} \;\bigg|\; 2 \text{ M RbCl} \;\bigg|\; 3.5 \text{ M KCl} \;\bigg\|\; AgCl \;\bigg|\; Ag.$$

The error of this method is not more than 1%, as it can be seen in Fig. 2.

REFERENCES

1. P.Browne, Meeting of ICSOBA, 1973, Nice, France, p.216
2. Farkas F., Gombos M., Klug O., Kovács F., Magyar Kém.Lapja
   <u>30</u>, 460 /1975/
3. D.J.G.Ives, G.J.Janz, Reference Electrodes. Academic Press,
   New York, /1961/, Chapter III and IV
4. H.S.Harned, B.B.Owen, The Physical Chemistry of Electrolytic
   Solution 3rd, Ed. Reinhold, New York, p.430, 1958
5. Handbook of Analytical Chemistry, Ed. L. Meites Mc.Graw-Hill,
   New York, 1963
6. R.P.Buck, Anal.Chem. <u>48</u>, 23 R /1976/

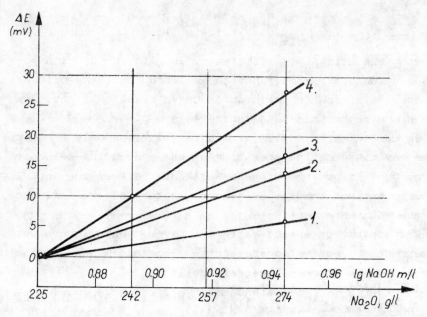

Figure 1. Calibration curves of a sodium ion-selective glass
electrode. 1/ Calculated electrode potential - concentration function if the activity coefficient 1
and without liquid-junction, slope 1,2 mV/10 g/1
Na_2O. 2/ Calculated with the application of the
literary activity data, slope 2,9 mV/10 g/1 Na_2O.
3/ Calculated with the measured mean activity
coefficients, slope 3,5 mV/10 g/1 Na_2O. 4/ Calculated as curve 3 and the liquid-junction is also
considered, slope 5,5 mV/10 g/1 Na_2O.

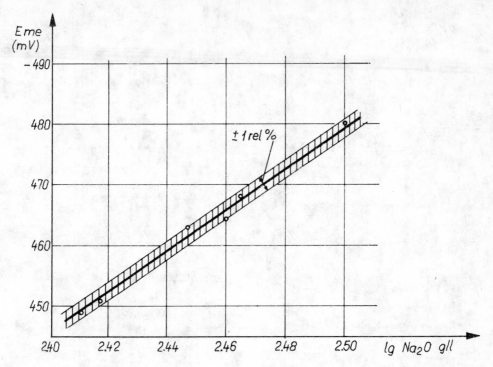

Figure 2. Determination of sodium content of different plant
liquors by sodium ion-selective electrode, slope
5,2 mV/10 g/l Na_2O.

QUESTION

Participants of the discussion: A.Hulanicki, L.Tomcsányi

Question:
Were your measurements performed at 25°C only or at higher
temperatures as well ?

Answer:
Satisfactory results have also been obtained with a cell
thermostated at 50°C.

ON THE VARIATION OF ELECTRICAL POTENTIAL
BETWEEN THE BULK OF GLASS ELECTRODE AND THE SOLUTION

Z. BOKSAY, G. BOUQUET AND M. VARGA

Department of General and Inorganic Chemistry, Eötvös Loránd University, Budapest, Hungary

ABSTRACT

The additive terms of the overall electrode potentials for a pH sensitive glass electrode and a sodium ion sensitive one, resp., have been compared. Potential values assigned to certain structural units within the electrical double layer have been estimated from the pH dependence of dissolution rate of the glass investigated. It has been concluded that every sodium glass electrode is probably covered with an extremely thin pH sensitive layer.

DISCUSSION

Every time when we try to extend the range of theoretical function of a glass electrode or to improve its response time, the question arises how the electrode works and what are the values of the terms contributing to the overall electrode potential.

The behaviour of a glass electrode is mostly determined by the glass components, first of all, by the type of alkali ion. In this paper we are dealing with two sodium glasses, one for pH measurement and one for sodium ion activity determination. The phase boundary potential of a pH sensitive electrode [1] is given by

$$E_p = E^o + \frac{RT}{F} \ln \frac{[H^+] \, h_H}{a_H} \tag{1}$$

where E^o is a constant, $[H^+]$ denotes the hydrogen ion activity in the solution, a_H that in the surface layer of glass phase and h_H is the proton acceptor concentration in the surface layer. The proton acceptor may be either an anionic group, $\equiv Si-O^-$, of the silicate network or a water molecule or anything else which can bind a proton.

a_H decreases with increasing pH and sodium ion activity in the solution due to the shifting of the ion exchange equilibrium

$$Na^+(aq) + \square(Na^+) + H^+(glass) \rightleftharpoons H^+(aq) + Na^+(glass) + \square(H^+) \tag{2}$$

where $\square(Na^+)$ is a vacancy in the glass structure able to accomodate a sodium ion and $\square(H^+)$ denotes the proton acceptor. The ion exchange equilibrium on the solution-glass interface is supposed to be instantaneously established after any change in the solution composition.

There is no overall equilibrium within the surface layer which is characterized by the so called profile curve showing the sodium ion relative concentration plotted against the depth. The profile curve of the McInnes-Dole glass after 14 days leaching in water at 40 ^{o}C is shown in Fig. 1. Owing to the concentration gradient, an interdiffusion of sodium ion and proton takes place in the layer. The inflexion point of the curve and the low alkali concentration in the outer part indicate a change in the structure. The dissolution of the silicate network should also be taken into consideration, which slows down the thickening of layer.

It follows from the uneven ion distribution in the surface layer and from interdiffusion that the local potential depends on the depth. The diffusion potential is quoted here in differential form

$$\frac{d\eta}{dx} = -\frac{RT}{F}\left(t_H \frac{d \ln a_H/h_H}{dx} + t_{Na} \frac{d \ln a_H/h_H}{dx}\right) \tag{3}$$

where η denotes the diffusion potential, x is the coordinate perpendicular to the surface, t_H and t_{Na} denote the transference number indicated. The gradient of the diffusion potential balances the fluxes, J_H and J_{Na}, of ions

192

$$J_H = t_H \varkappa \left(\frac{RT}{F^2} \frac{d \ln a_H/h_H}{dx} + \frac{1}{F} \frac{d\eta}{dx} \right) \tag{4}$$

$$J_{Na} = t_{Na} \varkappa \left(\frac{RT}{F^2} \frac{d \ln a_{Na}/h_{Na}}{dx} + \frac{1}{F} \frac{d\eta}{dx} \right) \tag{5}$$

which are regarded along with the conductivity, $\varkappa$, as experimental data. (The calculation of fluxes is based on equation $J=D \cdot dc/dx$, where D denotes the concentration dependent diffusivity). If we consider the gradient of $\ln a_H/h_H$ and that of $\ln a_{Na}/h_{Na}$, resp., as a single variable and take all equations into account, it will be evident that the data available — at least at the present stage of our investigation — are insufficient to calculate the diffusion potential. While the calculation can not rigorously be performed, some valuable information may be get from conductivity measurements. The conductivity and the relative concentration of sodium ion are shown as a function of depth in Fig. 2 (preliminary publication in ref. 2). It should be noted that, for methodical reason, during the measurement the specimens were in contact with an atmosphere of relatively low humidity instead of diluted solution.

The high conductivity in the inner part of the surface layer and the corresponding activation energy of low value indicate a predominant proton mechanism in the charge transportation. Therefore the transference number of the proton is practically equal to unity and thus a formula is easily obtained for the diffusion potential in this part of the layer by integration of equation 3 between the respective limits x_2 and x_3

$$\int_{x_2}^{x_3} \frac{d\eta}{dx} \, dx = \frac{RT}{F} \ln \frac{a_{H2} \, h_{H3}}{a_{H3} \, h_{H2}} \tag{6}$$

Considering the variation of the proton activity and that of proton acceptor concentration, we may conclude that the potential gradient is directed outwards. In the environment of the conductivity minimum, the sodium ion may also be involved into the electrical conduction the mechanism of which needs further investigations. Thus the problem of the diffusion potential in this range has not been resolved yet. The outer part of the surface layer shows a high conductivity especially when saturated with water. Since the diffusion rate is also high, the ion exchange equilibrium is supposed to be extended over the whole part with low alkali concentration.

When the ion exchange is shifted, a diffusion potential is formed in the outer part of surface layer owing to the initiated diffusion. The gradient of the latter

$$- \frac{d\eta}{dx} = \frac{RT}{F} (t_H \frac{d \ln a_H/h_H}{dx} + t_{Na} \frac{d \ln a_{Na}/h_{Na}}{dx}) \qquad (7)$$

transforms into that of eqilibrium potential as the two differentials in parentheses approach each other.

Recently we succeded in assigning a potential value to a structural unit in the electrical double layer on the phase boundary by using the data of dissolution measurements. The dissolution rate of the glass was found to increase with hydroxide ion concentration in the pH range over 5, however, not proportionally [3]. Fig. 3 shows the logarithm of the dissolution rate plotted against the pH of the solutions containing 0.2 mole Na$^+$ per litre. The slope indicating the apparent reaction order in the linear section is as high as 0.22. The fractional order of the reaction was explained by considering the effect of the varying phase potential on the movement of hydroxide ion. However, only a certain part of the phase potential is effective, namely the voltage between the solution bulk and the spot of the rate determining step of reaction (Fig. 4).

Since the hydroxide ion reacts with neutral groups of silicon oxygen tetrahedron more readily, than with anionic groups, $\equiv$Si-O$^-$, we may write

$$v = A [OH^-] [\equiv Si-OH] \exp(- \frac{\Delta H^\ddagger - FE_p\beta}{RT}) \qquad (8)$$

where v is the reaction rate, A the preexponential factor, $\Delta H^\ddagger$ the activation enthalpy and β the transfer coefficient. Inserting E_p given by equation 1 into equation 8 and considering

194

the ionic product of water we obtain

$$v = [\equiv Si-OH] [OH^-]^{1-\beta} \qquad (9)$$

in accordance with experiments.

An oxygen bridge in the silicate network is broken by the attack of a hydroxide ion as follows

$$\equiv Si-O-Si\equiv \quad \longrightarrow \quad \equiv Si \quad {}^-O-Si\equiv$$
$$OH^- \qquad\qquad\qquad OH \qquad\qquad (10)$$

After the breaking of the last oxygen bridge the silicon oxygen tetrahedron enters the solution as an orthosilicic acid molecule:

$$-O-\underset{\underset{O}{|}}{\overset{\overset{O}{|}}{Si}}-O-\underset{\underset{O}{|}}{\overset{\overset{OH}{|}}{Si}}-O- \rightarrow -O-\underset{\underset{OH}{|}}{\overset{\overset{OH}{|}}{Si}}-O-\underset{\underset{OH}{|}}{\overset{\overset{OH}{|}}{Si}}-O- \rightarrow -O-\underset{\underset{OH}{|}}{\overset{\overset{OH}{|}}{Si}}-OH \xrightarrow{\text{phase boundary}} H_4SiO_4 \qquad (11)$$

network forming tetrahedra   chain forming tetrahedra   terminal tetrahedron

The reaction rate was found to be influenced by increasing the sodium chloride concentration or adding aluminium chloride or other salts. Consequently, the terminal tetrahedra on the surface are involved into the rate determining step. The potential on their places is equal to $\beta E_p = 0.72 E_p$.

The other model glass, the Lengyel-Da glass, shows theoretical sodium function, provided that the ratio of sodium ion activity to hydrogen ion activity is higher than 10^2. Its surface layer, if any, is thinner than 30nm. The diffusion potential can practically be excluded. In spite of the profound differences between the two model glasses, a close similarity in the pH dependence of dissolution rate occurs according to Fig. 5 (from an unpublished paper).

Considering this similarity we are forced to assumed that a thin pH sensitive layer is formed on the surface of the Lengyel-Da glass. Since the sodium ion function of this glass is attributed to boron oxide and alumina content of its network, the latter in the pH sensitive thin layer is probably depleted in these components. While the potential in this

195

layer depends on pH, as it is shown schematically in Fig. 6, the electrode potential is, nevertheless, determined by pNa owing to an assumable equilibrium between the bulk of glass and the solution.

CONCLUSION

If the suggested mechanism of the dissolution applies to silicate glasses in general, we are lead to the paradox conclusion that every sodium sensitive glass electrode is covered by a pH sensitive layer.

REFERENCES

1. Z. Boksay and B. Csákvári, Acta Chim. Acad. Sci. Hung. 67, 157 (1971); Z. Boksay, in Ion-selective Electrodes (E. Pungor ed.), Akadémiai Kiadó, Budapest, 1975. P. 245.
2. Z. Boksay, Wiss. Ztschr. Friedrich-Schiller-Univ. Jena, Math.-Nat. R. 28, 477 (1979)
3. Z. Boksay and G. Bouquet. Physics Chem Glasses 21, 110 (1980)

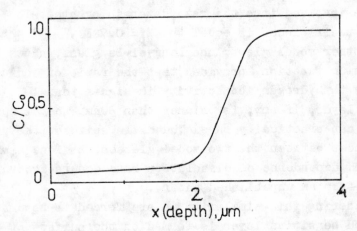

Figure 1. Relative concentration , C/C_o, of sodium ion in the surface layer of McInnes-Dole glass (22 % Na_2O, 6 % CaO, 72 % SiO_2) after 14 days leaching at 40 °C, plotted against the depth.

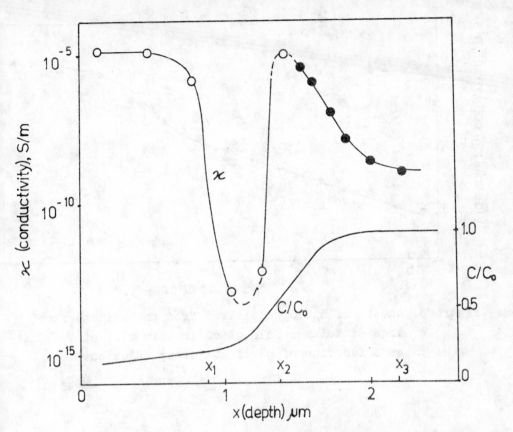

Figure 2. Variation of the conductivity and C/C_o with depth
in the surface layer of Mc^{I}nnes-Dole glass. The
electrical measurement was performed at 20 ^oC in
low humidity atmosphere (o   30 %, • ~0 %).

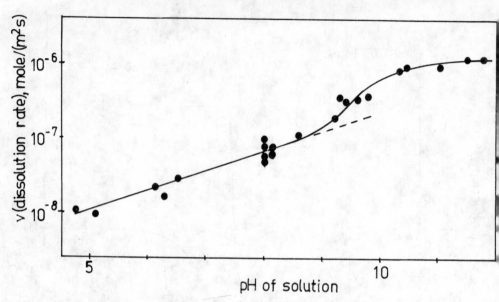

Figure 3. Amount of H_4SiO_4 dissolved from the unit surface area of McInnes-Dole glass in unit time at 75 ^oC as a function of pH of attacking solution.

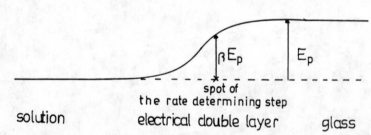

Figure 4. Spot of the rate determining step in the electrical double layer.

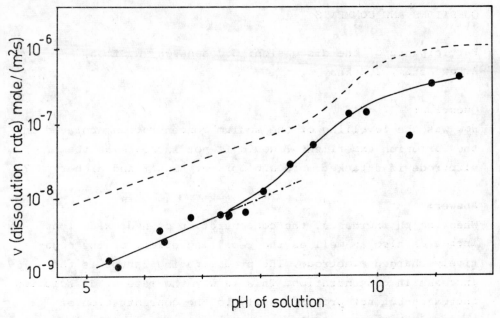

Figure 5. pH dependence of the dissolution rate for Lengyel-Da glass (11 % Na_2O, 11 % B_2O_3, 3 % Al_2O, 75 % SiO_2) compared to that of McInnes-Dole glass.

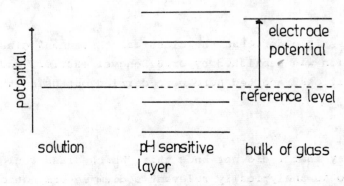

Figure 6. Schematic diagram showing potential levels in the pH sensitive layer of a sodium ion sensitive glass electrode, for some unspecified pH values, and the potential level in the bulk of glass and in the solution, resp. The electrode potential is fixed by an arbitrary constant activity of sodium ion.

QUESTIONS AND COMMENTS

Participants of the discussion: G.Johansson, W.Simon,
A.Lewenstam, Z.Boksay

Question:
How was the levelling of the sodium concentration achieved in
the corrosion experiment ? Would it not increase as the
electrode is attacked more and more at higher and higher pH ?

Answer:
When the pH increases, the concentration of hydroxide ions
increases also as well as the repelling effect of the negatively charged electrode. The predominating factor is the
increase in concentration. This is why the rate of dissolution
increases but not proportionally to the concentration of
hydroxide but more slowly.

Question:
In one of the slides the sodium-selective glass electrode
consisting of a boron aluminium silicate was shown. What
advantage does this have over the classical "Eisenman" glass?

Answer:
This glass is older than the classical "Eisenman" glass. The
composition was published by Dr.B.Lengyel earlier, and the
investigations reported here were carried out using this
glass.

Question:
I am sorry that I did not know this. Which glass would you
prefer to do analytically relevant sodium determinations
for example in blood serum ?

Answer:
I would suggest the Hungarian glass electrode produced by
Radelkis.

Question:
What is the difference in the dissolution rates of these two
types of glasses ? How can it be explained ?

Answer:
The constants in the equation are different, both the activation enthalpies and the transfer coefficients are different
for the two kinds of glasses.

The first difference may be due to the presence of some
alumina in the surface layer and of aluminium/III/ ions in
the adsorption layer, and also to the presence of boric acid.

The difference in the transfer coefficient may be considered
as a consequence of the difference in the layer thickness.
This is, however, rather an assumption than a proved theory,
but it is not important for the time being for continuing
this work.

METROLOGICAL BASIS OF CALIBRATION OF ION-SELECTIVE ELECTRODES AT ELEVATED TEMPERATURE

É. DEÁK

National Office of Measures, Budapest, Hungary

INTRODUCTION

It is well-kown that ion-selective electrodes respond to the activity and not the concentration of the measured ions. If we want to determine the ion activity in a solution of unknown composition by direct potentiometry we need a fix point, a reference solution with known ion activity to transform the measured potential difference to ion activity value. In case of the hydrogen ion activity this problem is well resolved by the pH scale used worldwide. With the common application of ion-selective electrodes the need arose for setting up activity scales also for other ions. Certainly, the base of the individual ion activity scales must be a convention.

Bates and Robinson /1/ have suggested that self-consistent scales of individual ionic activities might be based on the hydration theory of Stokes and Robinson and they have shown that in case of unassociated electrolytes activities of the individual ions can be derived from the mean activity coefficient γ_+, the osmotic coefficient $\mathscr{S}$, and from h, the sum of the hydration numbers for the ions of which the electrolyte is composed, provided the hydration number for a reference ionic species is assigned by convention. The basic assumptions of the convention are as follows: the reference ion is the chloride; its hydration number is zero; the hydration numbers of ions are independent from their concentration and they are additive. These principles and, consequently the recommended activity scales are not thermodynamically exact. Nevertheless they are very useful in practice. The results of different measurements /2,3/ provided strong confidence in these scales.

This conventional ion activity scale for alkali halide solutions has been defined only for 25°C, but ion-selective electrodes are often used in solutions of different temperatures, for instance in natural and waste waters, in biological fluids, etc. Therefore, to assure a common metrological basis for these measurements an attempt is made to extend the ion activity scales to the temperature range 10-60°C for sodium chloride and potassium chloride solutions. To define the ion activities at different temperatures the values were calculated according to the hydration theory, and the correctness of the results, i.e. of the applied assumptions was examined by measuring the emf of a cell with transference.

THEORETICAL

The first step of the calculation was the determination of the hydration numbers of the ions. For this purpose we needed the osmotic coefficients and mean activity coefficients in the given temperature range. To the extent possible experimental data were used and this fact determined the tested concentration and temperature range. The mean activity coefficients were measured by Harned,Nims and Cook /4,5/, the osmotic coefficients by Stoughton, Lietzke /6/ and Jákli /7/. The mean activity coefficients of four molal potassium chloride at 50°C and 60°C are extrapolated values. The hydration numbers for sodium chloride and potassium chloride were calculated according to the method of Robinson and Stokes. They are shown in the Table 1. The convention, that the hydration number of chloride is zero, was kept unchanged. Using these h values the following expressions have given the single ion activity coefficients:

$$\log \gamma_{M^+} = \log \gamma_{\pm} + 0.00782 \text{ hm} \varphi$$

$$\log \gamma_{Cl^-} = \log \gamma_{\pm} - 0.00782 \text{ hm} \varphi$$

where h is the hydration number of electrolyte, m the molality and φ the osmotic coefficient.

The variation of activity coefficients in sodium chloride solutions with concentration and temperature is shown on Fig.1. Knowing the single ion activity coefficients at different temperatures the pNa, pK and pCl values were calculated for the 0.5 mol kg^{-1}-4.0 mol kg^{-1} concentration range. The computed values are given in the Table 2. and 3. As it could be waited the differences are small but significant.

EXPERIMENTAL

The results of calculation were checked by measurements. The following two cells were measured in the given temperature range:

$$Ag|AgCl|KCl /x \text{ mol kg}^{-1}/ \vdots KCl /4 \text{ mol kg}^{-1}/|AgCl|Ag$$
$$Ag|AgCl|NaCl/x \text{ mol kg}^{-1}/ \vdots KCl /4 \text{ mol kg}^{-1}/|AgCl|Ag$$

Thermoelectrolytic type classical silver-silver chloride electrodes were applied. The electrodes were produced in our laboratory, their standard potential was determined by means of hydrogen electrode at 25°C. The potential difference between the applied electrodes was $\pm$ 30 microvolt maximum. Taking into consideration, that the given cells are symmetrical for the electrodes, we supposed, that the error of the electrode potentials does not cause mistake in our measurements.

In each case one of the half-cells contained four molal potassium chloride solution and the other one was filled with sodium chloride or potassium chloride solutions of different concentration. All solutions were saturated with silver chloride. The half-cells were connected by a bridge specially designed for this measurement /3/. The liquid boundary between the two solutions was situated in one arm of the U shape bridge and it should be regarded as continuous-mixture type junction. To compute the liquid junction potential the Henderson equation was applied, inserting in it the limiting ionic mobilities. In Table 4. we can see the variation of the liquid junction potential depending on the temperature.

Knowing the liquid junction potential and the pCl value of the solutions, which was defined above, the e.m.f. of the cells could be calculated. The e.m.f. of these cells was measured as well.

RESULTS

The comparison of the measured and computed data is summarized in Table 5. E_m results given here are mean values of three measurements. It is appearent, that there are slight differences between the measured values and those calculated on the basis of Bates-Robinson convention. The maximum value of the differences corresponds to about 0.02 pCl.

Data suggest, that the convention based on the hydration theory can be applied also in this broaden temperature range. Thus the possibility is given for definition of conventional single ion activity values for sodium chloride and potassium chloride solutions from $10^{\circ}C$ to $60^{\circ}C$. As a result of this work, presented here, we issue certified sodium and potassium chloride reference materials with assigned single ion activity values given in the Table 2.and 3.

REFERENCES

1. R.G.Bates and R.A.Robinson, Pure Appl.Chem.37,575/1974/
2. J.Bagg and G.A.Rechnitz, Anal.Chem.45,271./1973/
3. É.Deák,Conference .on ISEs 1977.Akadémiai Kiadó,Budapest
                                                    1978.
4. H.S.Harned and L.F.Nims, J.Am.Chem.Soc.54.423./1932/
5. H.S.Harned and M.A.Cook,J.Am.Chem.Soc.59,1290/1937/
6. R.W.Stoughton and M.H.Lietzke,J.Chem.Eng.Data 12,101/1967/
7. Gy.Jákli,Dissertation,Budapest,1973.

Table 1. Hydration numbers of NaCl and KCl
          depending on the temperature

°C	10°	20°	30°	40°	50°	60°
h_{NaCl}	3.4	3.5	3.6	3.7	3.7	3.8
h_{KCl}	1.9	1.9	2.0	2.1	2.1	2.2

Table 3. Computed pK and pCl values for KCl solutions

Molality	10°C		20°C		30°C		40°C	
	pK	pCl	pK	pCl	pK	pCl	pK	pCl
0.5	0.483	0.496	0.481	0.494	0.480	0.494	0.483	0.498
1	0.210	0.236	0.206	0.232	0.205	0.233	0.205	0.234
1.5	0.044	0.083	0.039	0.079	0.036	0.078	0.034	0.079
2	-0.078	-0.024	-0.086	-0.032	-0.092	-0.034	-0.093	-0.033
2.5	-0.177	-0.109	-0.187	-0.118	-0.193	-0.121	-0.196	-0.119
3	-0.263	-0.181	-0.272	-0.189	-0.279	-0.191	-0.282	-0.189
3.5	-0.339	-0.242	-0.350	-0.252	-0.357	-0.253	-0.361	-0.251
4	-0.408	-0.297	-0.417	-0.305	-0.426	-0.308	-0.432	-0.307

Table 2. Computed pNa and pCl values for NaCl solutions

Molality	10°C		20°C		30°C		40°C		50°C		60°C	
	pNa	pCl	pNa	pCl	pNa	pCl	pNa	pCl	pNa	pCl	pNa	pCl
0.5	0.458	0.483	0.457	0.482	0.456	0.482	0.456	0.483	0.458	0.485	0.461	0.488
1	0.163	0.212	0.159	0.210	0.156	0.209	0.155	0.210	0.156	0.210	0.156	0.212
1.5	-0.021	0.053	-0.029	0.049	-0.035	0.046	-0.038	0.045	-0.039	0.045	-0.038	0.048
2	-0.167	-0.064	-0.178	-0.070	-0.185	-0.074	-0.190	-0.075	-0.190	-0.075	-0.190	-0.072
2.5	-0.289	-0.156	-0.302	-0.164	-0.312	-0.168	-0.316	-0.168	-0.317	-0.168	-0.317	-0.164
3	-0.399	-0.234	-0.415	-0.244	-0.426	-0.248	-0.430	-0.248	-0.431	-0.247	-0.433	-0.243
3.5	-0.502	-0.284	-0.519	-0.313	-0.530	-0.316	-0.536	-0.315	-0.537	-0.316	-0.538	-0.312
4	-0.598	-0.364	-0.617	-0.374	-0.630	-0.377	-0.636	-0.377	-0.637	-0.375	-0.639	-0.370

Table 4. Liquid junction potentials,[*] mV

Junction: x mol kg^{-1} soln. ┆ 4 mol kg^{-1}KCl

Tempe-rature °C	Molality, NaCl							
	0.5	1	1.5	2	2.5	3	3.5	4
10°	-0.7	-1.6	-2.3	-2.9	-3.4	-3.8	-4.2	-4.5
20°	-0.3	-1.4	-2.1	-2.7	-3.2	-3.6	-4.0	-4.4
30°	-0.1	-1.2	-1.9	-2.6	-3.1	-3.5	-4.0	-4.3
40°	+0.2	-1.0	-1.8	-2.4	-2.9	-3.4	-3.8	-4.2
50°	+0.5	-0.7	-1.6	-2.2	-2.8	-3.3	-3.7	-4.1
60°	+0.8	-0.5	-1.4	-2.0	-2.6	-3.1	-3.5	-3.9

	Molality, KCl						
	0.5	1	1.5	2	2.5	3	3.5
10°	0.6	0.4	0.3	0.2	0.1	0.1	0.1
20°	0.9	0.6	0.4	0.3	0.2	0.1	0.1
30°	1.1	0.8	0.5	0.4	0.3	0.2	0.1
40°	1.3	0.9	0.6	0.4	0.3	0.2	0.1

[*]Positive ε_j signifies a boundary of polarity -┆+

Table 5. Comparison of the calculated and the measured e.m.f. values /in mV/

NaCl mol kg^{-1}	10°C		40°C		60°C	
	E_c	E_m	E_c	E_m	E_c	E_m
0.5	43.1	43.1	48.9	48.8	51.4	51.5
1	30.2	30.4	32.1	32.5	33.5	33.7
1.5	22.0	22.3	22.7	23.2	24.5	24.8
2	16.0	16.4	16.8	17.3	17.2	17.5
2.5	11.3	11.7	11.5	11.9	11.7	12.1
3	7.3	7.9	7.1	7.4	7.0	7.2
3.5	4.9	5.0	3.3	3.7	2.8	3.2
4	0.7	1.4	-0.2	+0.2	-0.6	-0.2

KCl mol kg^{-1}	10°C		40°C	
	E_c	E_m	E_c	E_m
0.5	44.0	44.2	48.7	48.9
1	30.0	30.4	32.7	32.8
1.5	21.1	21.4	23.4	23.6
2	15.1	16.1	16.6	17.4
2.5	10.5	10.8	11.4	11.8
3	6.4	6.7	7.1	7.6
3.5	3.0	3.2	3.4	3.5

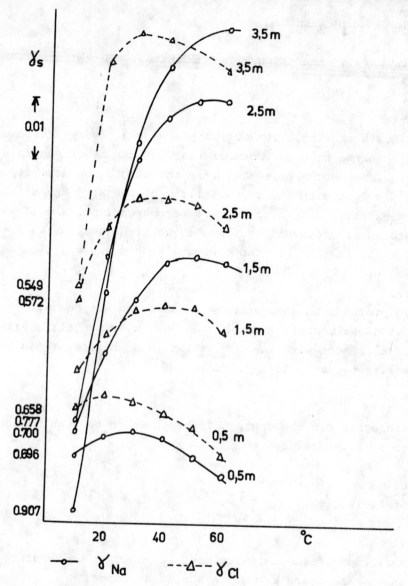

Figure 1. Variation of the single ion activity coefficients in NaCl solutions depending on the concentration and temperature

QUESTIONS AND COMMENTS

Participants of the discussion: R.G.Bates, J.D.R.Thomas,
É.Deák

Comment:
These tests on single ion activity scales are very interesting
but I think we have to remember that one can never separate
the liquid junction potential from the single ion activity and
so we are left with the impossibility of checking any of these
scales exactly. Such a scale must be conventional and if we
have to adopt a conventional scale, we might just as well not
worry about testing it, and everyone should use the same scale.

Answer:
In my opinion such a conventional scale can not be set up
without experimental studies, and our experimental results
are in agreement with the convention, and therefore this
convention seems to be applicable.

Comment:
I quite agree,but the scale must be shown to be reasonable
and this is important.

Answer:
Yes, this is true.

Question:
The variation of the activity coefficient with temperature
was shown, and there was quite a sudden change, then a
maximum or minimum was reached, and the curve came back again.
Could you give an explanation for this levelling out ?

Answer:
This variation depends on the variation of the mean activity
coefficients with temperature, but I can not give any further
explanation.

Comment:
According to our experience, these all follow a Hückel type
of curve, they come down and at higher concentrations they
rise again because of the nature of the convention.

Question:
Is this kind of apparent increase based on modified Debye-
-Hückel calculations of the activity coefficient, or is it a
true one or an imaginary one ?

Answer:
These are classical electrochemical data, the single and mean
activity coefficients as well. This behaviour does not depend
on the ion-selective electrode characteristics.

ALL-SOLID STATE ION-SELECTIVE ELECTRODES WITH INTEGRAL AMPLIFIER ELECTRONICS

T. A. FJELDLY AND K. NAGY*

Electronics Research Laboratory, The University of Trondheim,
The Norwegian Institute of Technology, N-7034 Trondheim, Norway
*Division of Applied Chemistry, SINTEF, The University of Trondheim, The Norwegian Institute
of Technology, N-7034 Trondheim, Norway

ABSTRACT

All-solid state ion-selective electrodes based on silver-salt and LaF_3 membranes have been prepared and tested. Amplifier electronics has been integrated in the sensor by means of thick-film hybrid technology. This report contains construction details and test results, with particular emphasis on the fluoride electrode equipped with a newly developed reversible solid state membrane contact. Also included is a brief discussion of a solid state differential electrode based on similar technology.

INTRODUCTION

With the recent introduction of cost-effective microcomputers for limited and decentralized control purposes, the limiting factors in control system development and cost have gradually shifted to peripherals like sensors and actuators [1]. Increasingly, therefore, highly efficient manufacturing techniques from the microelectronic industry are adopted by sensor producers, a development which is further encouraged by the need for integral signal-conditioning and transmission electronics at the sensor site.

In ion-selective sensors, this has led to developments like the ISFET (Ion-Sensitive Field Effective Transistor) [2], the ChemFET (Chemically Sensitive Field Effect Transistor) [3] and the present all-solid state ion-selective electrodes with integral electronics [4-6]. In conventional electrodes, the chemically sensitive membrane is connected by an internal solution contact to a shielded cable for direct transmission of the high-impedance signal. However, with the introduction of impedance-transforming electronics in close proximity to the membrane, noise pickup in the transmission line is eliminated, and the signal level can be adjusted in accordance with the requirements of the processor.

In the present electrodes the impedance transformation is performed within the sensor body, just behind the membrane [5,6]. Important in this construction is the use of solid membrane contacts, which is desirable in terms of miniaturization, electrode ruggedness and compatibility with microelectronic manufacturing processes. Also, in order to achieve the desired electrode quality in terms of sensitivity, response time and stability, it is essential that the membrane contacts be reversible [7,8]. Silver metal is such a contact when used in conjunction with silver salt membranes [9]. A special solid contact has also been developed for use in fluoride electrodes [8]. It is prepared as a layered structure of LaF_3, AgF and Ag (metal). The fused LaF_3-AgF interface allows the reversible exchange of F^- ions between the two materials, and reversible contact to the external wiring is provided by the silver metal on top of the AgF.

This report contains constructional and experimental details of these electrodes, with particular emphasis on the fluoride electrode. Preliminary experience with a novel integrated differential electrode using two "orthogonal" membranes is also briefly discussed.

ELECTRODE CONSTRUCTION

The integral impedance transforming electronics consists of an operational amplifier chip (RCA CA 3140) connected in voltage follower configuration. It is located just behind the membrane on a miniature ceramic substrate where also the necessary connecting lines are printed with thick-film gold paste. The whole construction is sealed and potted inside a PVC tube, as shown in Fig. 1. Only three external connections are necessary, two for carrying the + and - 5V dc voltage from the power supply, and one output signal line. For additional noise immunity, an extra line is normally used for connecting external ground to a thick-film metal shield covering the back-side of the ceramic substrate.

As already noted, the solid contact between the ion-conducting membrane and the electronic circuit is of primary importance. Reversibility in the contact assures rapid establishment of thermodynamic equilibrium at the interface, thus preventing long-term potential drift [7,8]. Non-reversible, blocked interfaces are also applicable [10], but only when the interfacial relaxation processes proceed sufficiently slowly not to interfere with the measurements [8]. Examples of acceptable blocked interfaces can be found in MOSFET (metal oxide semiconductor transistor) devices, such as the high-impedance input stage of the operational amplifier used here.

Reversible contacts are normally achieved in solid junctions between ionic and electronic conductors when a parent metal M is contacted with solid electrolytes of the form MX. The reversible reaction taking place at the interface is then

$$MX + e^- \rightleftarrows M + X^- \tag{1}$$

Other metals can also be used provided their salt formation free energies are positive relative to the parent metal [9]. Thus, the construction of proper contacts is straightforward in the case of silver salt membranes, where silver metal is conveniently used [5,6,9]. A number of silver salt membrane electrodes based on the present electrode design have been constructed and tested. The silver contacts were applied either by thin-film vacuum deposition or as silver conducting epoxy.

In the case of the fluoride-selective LaF_3 membrane, the lanthanum metal does not lend itself easily as a contact material because of its reactivity

216

in air. Instead, a sandwich structure with Ag, AgF and LaF$_3$ was constructed, as indicated schematically in Fig. 2 [8], allowing reversible reactions
to take place at both the LaF$_3$-AgF and at the Ag-AgF interfaces, as indicated. The particular advantage of this contact is that the intermediate
AgF layer itself is a Ag$^+$ ionic conductor [11], assuring a reasonably low
overall ohmic resistance in the system.

Satisfactory contacts were made as fused junctions by melting small amounts
of AgF at the LaF$_3$ surface, at a temperature of 450 - 500°C. In order to
avoid excessive decomposition of AgF, the melting was performed in a dry,
inert atmosphere. Nonetheless, some decomposition of AgF took place,
probably leaving metallic silver dispersed throughout the AgF. The contacts were completed by applying silver conducting paint or epoxy on top
of the AgF.

In practical analytical situations, full advantage of the present all-solid
state technology can be derived only when the classical liquid-junction
reference electrode is eliminated. For a number of analytical routines
this can be done by working in a differential mode, using two "orthogonal"
membranes [12]. By using a suitable buffer solution, the surface potential
barrier of one of the membranes can be fixed, allowing it to act as a
reference element for the other. Using the basic all-solid state design
discussed above, a fluoride (LaF$_3$) and a sulphide (Ag$_2$S) membrane were
combined in the same sensor body, wherein also the differential amplifier
circuit, shown in Fig. 3, was integrated. A grounded platinum electrode,
also contained in the sensor unit, served to stabilize the potential at the
electrolyte to within the working range of the amplifier.

EXPERIMENTAL RESULTS

Electrodes with Silver Salt Membranes

For the purposes of testing the basic electrode design, a number of
electrodes with pressed pellet silver salt membranes were constructed. In
experimental tests of sensitivity, selectivity, response time and stabili-
ty, electrodes for measuring ions such as Cℓ^-, I$^-$, CN$^-$, S^{2-}, SO$_4^{2-}$, Ag$^+$ and
Pb^{2+} were found to behave very satisfactorily. This work has already been
documented elsewhere [5,6,13].

Fluoride Electrodes

The properties of the novel solid-state contact to LaF$_3$ were tested
both electrically and by incorporation in ion-selective electrodes. From
frequency dependent admittance measurements and low-frequency current-
voltage characteristics, an equivalent circuit for the Ag-AgF-LaF$_3$ contact
was derived, showing a predominantly ohmic, non-blocked behaviour in the
frequency range of analytical interest [8]. In a similar test, a direct
silver metal contact on LaF$_3$ [4] was shown to have a blocked, capacitive
behaviour [8]. In terms of applicability in potentiometric analysis, this
difference turns out to be of vital importance, as will become evident.

In Fig. 4 are shown typical response curves obtained with the present
electrodes using the Ag-AgF-LaF$_3$ membrane structure. The measurements
were performed by pipetting known additions of NaF standard into 50 ml of
stirred buffer solutions, either TISAB or 1M HCℓ. The curves are seen to
be in close agreement with theory, both with regard to slope and detection

limit. The slightly high detection limit in TISAB can be explained in terms of the complexing power of TISAB, which causes excessive dissolution of membrane material [8,14].

Of particular interest is the response time of this electrode, which was measured both with an injection technique [5], and in a regular analytical setup. With the injection technique, sample solution was rapidly injected into a narrow space between the ion-selective electrode and a reference electrode (Orion 96-01), suddenly replacing a small drop of initial solution bridging the gap. With this technique, time constants in the range 20 - 100 msec were found, depending on the experimental parameters. This was in reasonable agreement with similar measurements performed on a commercial electrode (Orion, Model 94-09).

In the regular analytical setup, stirred sample solutions (50 ml 1M HCl) were monitored under successive additions of known quantities of NaF. The time response was recorded on an x-t recorder, or with a computerized analytical system [15]. Typical responses for fluoride electrodes with the $Ag - AgF - LaF_3$ membrane system are shown in Fig. 5. For the highest concentrations, potential stability was achieved within a few seconds. A steady increase in response time could be observed with lower concentrations, below 10^{-4} M NaF. In comparison, LaF_3 membranes with a direct silver metal contact had response times on the order of minutes, and in addition, a long-term potential drift on the order of millivolts per minute was observed [8]. In a careful comparison the solid-state $Ag - AgF-LaF_3$ electrode and a commercial electrode (Orion 96-09) were found to be about equally fast in sample concentrations of normal analytical interest. Further details on these and other response time measurements are presented elsewhere [16].

Differential Electrodes

Preliminary investigations have been done on a differential potentiometric cell system, as described above. In Fig. 4 is shown the excellent fluoride response curve obtained with the differential electrode. The composition of the buffer solution was 1M CH_3COOH, 1M CH_3COONa and 10^{-3}M $AgNO_3$, designed to keep constant pH and the Ag^+ concentration. Known additions of NaF were added while stirring the buffer solution. Equivalent results were also obtained in the reverse mode, i.e. with a fixed fluoride level (10^{-3}M NaF in the same pH buffer) and varying the Ag^+ concentration. Response time and stability was about the same as for regular systems with a proper reference electrode. Owing to the integration of amplifier electronics in the sensor body, no significant increase in the noise level of the potential readings could be detected. A more comprehensive discussion of the differential solid state electrode will be presented elsewhere [17].

SUMMARY

Ion-selective electrodes with integral amplifier electronics, based on silver salt and LaF_3 membranes, have been prepared and tested with very satisfactory results. In particular, the fluoride electrode with a newly developed solid-state $Ag - AgF - LaF_3$ contact has been shown to compare favourably with conventional electrodes with internal filling solution, both with regard to sensitivity and response time.

An integrated differential electrode of the same basic solid-state design, but with two "orthogonal" membranes and an auxiliary platinum electrode, has also been constructed. Preliminary tests indicate that for special analytical situations, this electrode may be fully compatible with systems using the classical liquid-junction reference electrodes.

REFERENCES

[1]   W.G. Wolber and K.D. Wise, IEEE Trans. Electron Devices, ED-26, 1864 (1979).

[2]   P. Bergveld, IEEE Trans. Biomed. Eng., 17, 70 (1970) and ibid., 19, 342 (1972).

[3]   J.N. Zemel, Anal. Chem., 47, 255A (1975).

[4]   K. Nagy, T.A. Fjeldly and J.S. Johannessen, in Ion-Selective Electrodes, E. Pungor, Editor, p. 491, Akadémia Kiado, Budapest (1978).

[5]   T.A. Fjeldly and K. Nagy, SINTEF report STF 21 79009 (1979).

[6]   T.A. Fjeldly, K. Nagy and J.S. Johannessen, J. Electrochem. Soc., 126, 793 (1979).

[7]   R.G. Kelly, Electrochim. Acta, 22, 1 (1977).

[8]   T.A. Fjeldly and K. Nagy, J. Electrochem. Soc., 127, 1299 (1980).

[9]   R.P. Buck and V.R. Shepard, Jr., Anal. Chem., 46, 2097 (1974).

[10] R.P. Buck, in Ion-Selective Electrodes in Analytical Chemistry, Vol. 1, H. Freiser, Editor, Plenum Press, New York (1978).

[11] A.M. Raaen, I. Svare and T.A. Fjeldly, Phys. Rev. B21, 4895 (1980).

[12] M.J.D. Brand and G.A. Rechnitz, Anal. Chem., 44, 616 (1970).

[13] K. Nagy, E. Kleven, T.A. Fjeldly and D. Fremstad, Z. Anal. Chem., 295, 362 (1979).

[14] M.S. Frant and J.W. Ross, Anal. Chem., 40(7), 1169 (1968).

[15] K. Nagy and E. Keul, Proc. Meeting of the Metallurg. Soc. of AME – Denver 1978, paper no. A78-39.

[16] K. Nagy and T.A. Fjeldly, this conference.

[17] T.A. Fjeldly and K. Nagy, unpublished.

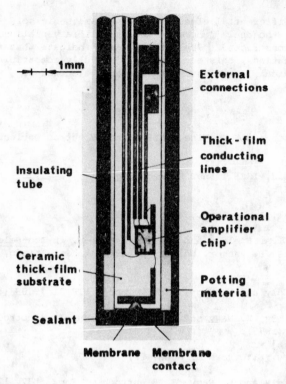

Fig. 1   Basic design of the all-solid state ion-selective electrode with integral amplifier electronics.

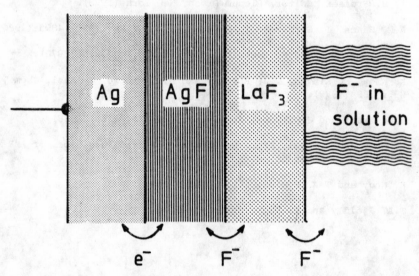

Fig. 2   Schematic structure of the reversible solid contact on LaF_3.

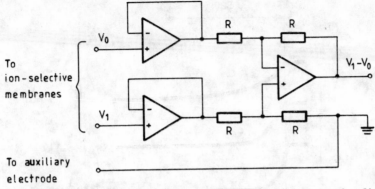

Fig. 3 Amplifier circuit with high-impedance input used in the differential electrode. Operational amplifiers: RCA CA 3240 and RCA CA 3140. Four identical resistors: 1 kΩ.

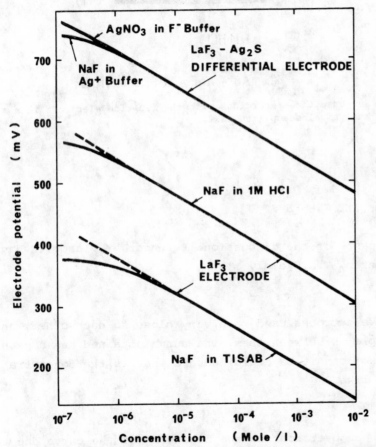

Fig. 4 Response curves obtained with the fluoride electrode and with the differential $LaF_3 - Ag_2S$ electrode (see text). The slope of the response curve in the Ag^+ detecting mode is inverted by the differential amplifier.

221

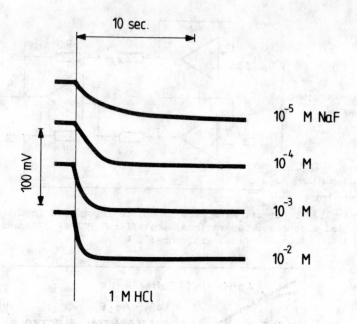

Fig. 5   Response time recordings with the fluoride electrode in a regular analytical setup (see text).

QUESTIONS AND COMMENTS

Participants of the discussion: E.Pungor, Y.Vlasov, K.Nagy, A.Lewenstam, T.A.Fjeldly

Question:
The two electrodes used are very close to each other, and the sample is between them. The curve obtained has a peculiar shape. How can it be explained and what is the concentration profile like ?

Answer:
There is an overshoot at the beginning of the curve due to the short-circuit between the trigger electrodes and the measuring electrodes.

Question:
In the device described a solid contact was applied. How
long is the lifetime of this contact and how it prepared ?

Answer:
We have been using this electrode for two years and it is
still working quite well, without any problem.

Silver chloride is known to be hygroscopic and when we
make a contact we immediately seal it with silver-conducting
epoxy resin and we put the whole thing inside the electrode
body to protect it from water. This is the main reason for the
long lifetime of the electrode.

Question:
What sort of reference electrode was used ? Was the reference
electrode chosen according to the properties of the medium ?

Answer:
Different reference electrodes were used, the Radiometer
calomel electrode and the Orion double junction reference
electrode, and they performed properly.

Question:
What was the construction of your combination electrode ?

Answer:
It incorporated two selective membranes and the analyte contained the ion at a constant level to which one of the membranes
was reversible.

Question:
Was the surface layer of the fluoride ion-selective electrode
studied after using it in TISAB solution, possibly with the
purpose of drawing conclusion for the changes in response
time ?

Answer:

No such studies were made so far. At the earlier stage of the work we tried to do X-ray analysis of the LaF_3 surface after soaking in solutions with different compositions, and the results were reported at the 1977 conference in Budapest, but I think the effect of TISAB was not investigated.

SIMULTANEOUS DETERMINATION OF TETRAHYDRO-THIOPHENE, VOLATILE THIOLS AND HYDROGEN SULPHIDE USING ION-SELECTIVE ELECTRODES

T. GARAI,* T. BÁLINT** and M. SZŰCS***

*Laboratory of Inorganic Chemistry of the Hungarian Academy of Sciences, Budapest

**Research Institute of the Oil and Gas Industry, Veszprém
***Budapest Gasworks, Budapest

ABSTRACT

An electroanalytical method was developed for the simultaneous determination of tetrahydro-thiophene and volatile
thiols used as odorants of natural gas, in the presence of
hydrogen sulphide. Hydrogen sulphide and volatile thiols
were selectively precipitated in suitable absorbent solutions,
subsequently analyzed while tetrahydro-thiophene was catalytically decomposed to hydrogen sulphide and absorbed in an
alkaline solution of mercuric ions. The absorption was monitored potentiometrically using a sulphide selective electrode.

INTRODUCTION

The addition of odorants to natural gas in city gas
supply systems is a very important safety measure. The smell
permits the detection of any leak in the pipeline or in the
house and thus explosions and serious health hazards can be
avoided.

The sense of smelling is one of the most sensitive
instrument we have. However, it is subjective and it does not
provide quantitative data, thus quantitative analytical
methods are required for the determination of the odorant
content of gas pipeline systems.

The most widely used odorant are volatile thiols such as
ethanethiol or a mixture of low molecular weight thiols. These

compounds are very reactive and thus they easily loose their efficiency. In order to improve the odoration of natural gas a new odorant composition has recently been tested in Hungary, namely a mixture of ethanethiol and tetrahydro-thiophene. The smell of the latter compound is as unpleasant as that of the former, however, it is less volatile and rather stable. The gas contains about 0.3 /umol/l of each compound.

EXPERIMENTAL

Ion-selective electrodes were found to be suitable for the determination of volatile thiols [1,2] and they were also promising for the simultaneous analysis of hydrogen sulphide, ethanethiol and tetrahydro-thiophene content of natural gas. Thus expensive gas-chromatographic equipments can be replaced by an ion-selective electrode containing cell which is also advantageous from another point of view, namely it can easily be built into a portable instrument for the field test as well.

Tetrahydro-thiophene was found to be totally inactive from an electrochemical point of view. So, it had to be transformed into a detectable species. Thus the tetrahydro-thiophene content of the gas was thermally and catalytically decomposed in a micro reactor whereby it was transformed into hydrogen sulphide and the latter could be determined potentiometrically. Papp and Havas [3] suggested the use of silver sulphide based sulphide selective electrode as the indicator electrode for the mercurimetric titration of sulphide. In the present case the hydrogen sulphide containing gas had to be used as the titrant of known amount of reagent. By measuring the volume of the gas introduced until the equivalence point of titration the hydrogen sulphide content of the gas can be evaluated. Fig 1. represents the potentiometric titration curve recorded in the titration of 1 /umole Hg^{++} / the concentration of the latter beeing 1×10^{-4} mol/l approximat-

ely/ in 1 mol/l NaOH using a solution of Na_2S as the titrant.
The potentiometric cell consisted of a sulphide selective
electrode /Radelkis OP-S-7111/ and an 1 N calomel electrode
connected to the cell by a salt bridge containing 1 mol/l
KNO_3. An automatic burette /Radelkis type OP-930/ served for
the addition of the titrant. The e.m.f. of the cell was monitored by a high input impedance mV-meter and the output of the
latter was connected to an X-Y recorder /Hewlett Packard type
7000 AM/. The time axis of the recorder was synchronized with
the burette. It is apparent in Fig.1 that the potentiometric
curve is well-defined and a large potential step can be observed at the equivalence point even in such dilute solutions.

The sulphide content of the titrant was simultaneously
determined with the conventional iodometric procedure [4].
The results of the mercurimetric and iodometric titrations
were found to agree with an accuracy of 1 per cent.

The thermal and catalytical decomposition of tetrahydro-
thiophene was studied in the apparatus shown in Fig. 2. Natural gas containing mainly methane was pumped through the system
at a rate of 25 to 30 l/h. A wash bottle containing silver
oxide in alkaline isopropanol-water mixture was inserted to
remove the sulphur containing components of the gas. Various
amounts of a tetrahydro-thiophene solution in n-octane containing 0.1 /umole THT in 1 /ulitre were injected in the gas
stream by means of a microsyringe. The gas stream was conducted through the microreactor which consisted of a glass tube
30 mm long having 5 to 7 mm diameter containing a platinum
filament heating and cobalt-molybdenum type desulphurization
catalyst. The hydrogen sulphide content of the gas was absorbed in 1 mol/l NaOH containing a known amount of Hg^{++} ions.
The reaction of Hg^{++} with hydrogen sulphide was followed by
recording the e.m.f. of a sulphide sensitive electrode containing cell as a function of the volume of the gas.

The olefinic hydrocarbons present in natural gas or formed in the microreactor also reacted with mercuric ions. This interference, however, could be reduced by adding 0.1 mol/l KCl to the electrolyte containing Hg^{++} ions and by absorbing olefinic hydrocarbons in 75 per cent H_2SO_4 saturated with Na_2SO_4 [5]. The potentiometric titration curve was not affected by the presence of potassium chloride.

The percentage of the conversion to hydrogen sulphide of tetrahydrothiophene is shown in Fig. 3. In the case of thermal decomposition /heated platinum wire/, it is apparent that a heating power of 50 W was necessary for the total conversion. The filament temperature was approximately 1200 °C in this case. However, the decomposition of tetrahydro-thiophene to hydrogen sulphide was found to be practically complete with approximately 25 W heating power if thermal and catalytical conversion were combined.

The analysis of the gas mixture was performed in the apparatus shown in Fig. 4. A known amount of gas, generally 5 to 10 litres, was conducted through the system at a constant rate. The rate of the gas flow was controlled by a float type flow-meter /1/. Hydrogen sulphide and thiols were selectively absorbed in different solutions while tetrahydro-thiophene was decomposed in the microreactor prior absorption.

The first absorber /2/ contained 5 per cent cadmium acetate solution in dilute hydrochloric acid having a pH of approximately 1.2 which enabled the quantitative absorption of hydrogen sulphide. The hydrogen sulphide content of the solution could be determined by the conventional iodometric procedure after completion of the analysis [4]. The hydrogen sulphide concentration could be evaluated knowing the volume of the gas conducted through the system.

The second absorber /3/ contained a known volume of 1×10^{-5} mol/l silver nitrate solution in ethanolic ammonia

and ammonium nitrate /0.25 mol/l NH_3 + 0.1 mol/l NH_4NO_3/.
Ethanethiol is quantitatively precipitated in the solution.
The ethanethiol content of the solution was separately determined by potentiometric back-titration of the excess silver
nitrate in the absorbent. The titration error due to the adsorption of silver ions on the silver mercaptide precipitate
could be reduced below the measurement error by adding to an
aliquot part of the absorbent an amount of $1x10^{-3}$ mol/l KJ
solution equivalent to the initial silver ion concentration
of the solution. Potassium iodide reacted with the excess
silver nitrate of the absorbent. Since the solubility product
of silver mercaptide was found to be lower than that of
silver iodide, the amount of unreacted iodide was practically
equivalent to the ethanethiol absorbed in the solution. The
unreacted iodide content of the solution was potentiometrically titrated using $1x10^{-3}$ mol/l $AgNO_3$ solution as the titrant
and an iodide sensitive electrode /Radelkis OP-J-7111/as the
indicator electrode. The potentiometric titration curve was
found to be well defined even in such a dilute solution [1,2].

A potential step of 100 to 120 mV was observed as shown
in Fig. 5. The reference electrode was an 1 N calomel electrode connected to the solution by a suitable salt bridge.

After the removal of hydrogen sulphide and ethanethiol
the gas mixture was introduced in the third washing bottle
/Fig 4 /4// containing sulphuric acid in order to bind the
eventually escaping ammonia and olefinic hydrocarbons.

Tetrahydro-thiophene was practically not absorbed in the
above mentioned solutions. Thus the tetrahydro-thiophene
content of the gas was decomposed in the microreactor and the
hydrogen sulphide formed was absorbed in the potentiometric
cell in 1 mol/l NaOH containing 1 /umole mercuric ion as in
the model experiment shown in Fig. 1. The potentiometric cell
consisted of a sulphide sensitive electrode as the indicator
electrode and a Ag/AgCl reference electrode connected to the

bulk of the solution by means of a 1 mol/l KNO_3 salt bridge. The e.m.f. of the cell was recorded in the apparatus mentioned above. The potentiometric titration curve corresponded to that obtained with model solutions as shown in Fig. 1. The analysis can conveniently be carried out in the 0.1 to 1 /umol/l concentration range of each sulphur containing component with a standard deviation of 5 per cent.

REFERENCES

1. Garai,T., Szücs, M., Dévay J.: Conference on Coulometric Analysis, Mátrafüred, 1978. Akadémiai Kiadó, Budapest, 1978, 211.
2. Garai, T., Szücs, M., Dévay, J.: Hung. Sci. Instr. 46. 17 /1979/.
3. Papp, J., Havas, J.: Hung. Sci. Instr. 17. 17 /1970/.
4. Erdey, L.: Bevezetés a kémiai analizisbe II. Tankönyv-kiadó, Budapest, 1952.
5. Polgár, A., Jungnickel, J.L.: Determination of Olefinic Unsaturation. Organic Analysis Vol. 3. Interscience, New York, 1956.

Table 1

Analysis data of sulphur containing compounds injected in natural gas as the carrier

H_2S			EtSH			THT		
weighed in /umol	found /umol	dev. %	weighed in /umol	found /umol	dev. %	weighed in /umol	found /umol	dev. %
1.12	1.05	7	0.60	0.57	5	1.17	1.10	7
0.56	0.51	10	0.30	0.28	6	0.60	0.58	3
1.0	0.9	10	0.50	0.48	4	1.17	1.12	5

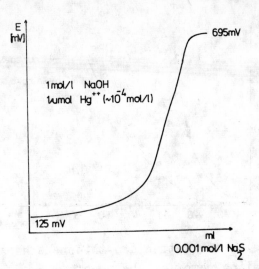

Fig. 1. Potentiometric titration curve of 1 μmole Hg^{++} /1x10^{-4} mol/1/ in 1 mol/1 NaOH using a Na$_2$S solution as the titrant.

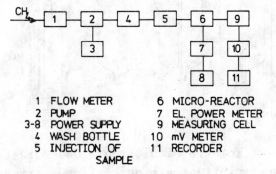

1 FLOW METER	6 MICRO-REACTOR
2 PUMP	7 EL. POWER METER
3-8 POWER SUPPLY	9 MEASURING CELL
4 WASH BOTTLE	10 mV METER
5 INJECTION OF	11 RECORDER
SAMPLE	

Fig. 2. Block diagram of the experimental setup.

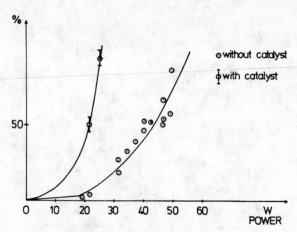

Fig. 3. Conversion of tetrahydro-thiophene as a function of the heating power of the reactor.

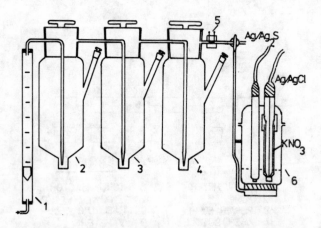

Fig. 4. Apparatus for the simultaneous determination of hydrogen sulphide, ethanethiol and tetrahydro-thiophene.

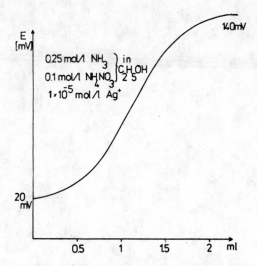

Fig. 5. Potentiometric titration curve recorded in the
determination of ethanethiol content of natural gas.

QUESTION

Participants of the discussion: E.Pungor, T.Garai

Question:
I have observed two equivalence points on the titration curve
shown in your figures. Could you comment on this phenomenon ?

Answer:
There is some kind of second inflexion point near the equivalence point of the titration curve indeed. The reason for
this phenomenon is not yet clear. It is probably connected
with the time lag of the electrode response at very low
sulphide concentrations.

ANODIC BEHAVIOUR OF SYNTHETIC COPPER(II) SULPHIDE

E. GHALI* and A. LEWENSTAM**

*Department of Mining and Metallurgy, Laval University Quebec, Canada, G1K 7P4
**Department of Chemistry, University of Warsaw, 02-093 Warsaw, Poland

INTRODUCTION

Copper (II) sulphide is one of the sulphides successfully used in variety of electrochemical processes and devices
such as electrowinning of copper and sulphur, solar energy
collector panels, high energy density batteries and – last
but not least – copper solid-state ion-selective electrodes
(ISE).

To gain more insight into the electrochemistry of copper
(II) sulphide and its role in the functioning of copper ISE
studies of the anodic behaviour of CuS in acidic chloride
medium were undertaken [1,2].

EXPERIMENTAL

Copper (II) sulphide was synthesized by distillation of
sulphur over prereduced copper under vacuum for 48 hours at
$450°C$, followed by cooling at the rate $25°C$ per day. This
was done in the sealed quartz-pyrex ampule using two independent furnaces [2]. The procedure applied enabled to produce pure CuS as proved with X-rays diffraction analysis.
Copper (II) sulphide supplied by Ventron with approx. 3% by
weight of Cu_9S_5 was also used [1]. The membranes of 82-96%
relative density (assuming $d_{CuS} = 4.67$ g/cm^3) were compres-

sed by means of hydrostatic and mechanical techniques. The
membranes were provided with solid electrical contact and
embedded in epoxy resin. The surface of the electrodes was
polished on emery papers /240-600 grit/ and on a diamond paste /1 μ size/.

RESULTS

Both types of the electrodes gave Nernstian response towards copper (II) ions with standard potential approximately
+280 mV vs SCE for Ventron and +345 mV for home made CuS.
The corrosion potential in 1 mol/1 hydrochloric acid saturated with argon was in the range +250 mV vs SCE for these two
types of electrodes [3].

The potentiodynamic studies in 1 mol/1 hydrochloric acid
presaturated with argon at 25°C were chosen to establish the
influence of various parameters on the anodic behaviour. The
potential-sweep curves with different scan rates are shown
in Fig. 1. Each curve represents first half-cycle starting
at the rest potential. The parabolic shape of voltammetric
curves with linear correlation of the peak current density
and square root of the scan rate indicate the active-passive
behaviour as a result of irreversible anodic reaction. The
successive voltammetric curves /Fig. 2/ show a suppression of
the current in each subsequent cycle. The presence of elemental sulphur - even after first half cycle - easily observed under scanning electron microscope led to conclusion
that the presence of the barrier-like film of sulphur impo-
sing transport limitations in the vicinity of the electrode
surface is primary responsible for active-passive behaviour
observed [1]. Although the purity of CuS is influencing the
magnitude of the anodic current, it does not alter the anodic behaviour described above, /Fig. 3/. The same relates to
the porosity of membranes, rotation of the electrode /up to
2500 rpm/, temperature increase /to 80°C/, oxygen presence
and $CuCl_2$ concentration changes. The only important factor,
which influences strongly the overall anodic process is the

236

total chloride and hydrogen ions concentration /Fig. 4/. The striking feature is that active-passive behaviour disappears in 10^{-2} mol/l HCl, while in 1 mol/l KCl at pH=2 this type of behaviour dominates and is accompanied with the presence of elemental sulphur as was evidenced with scanning electron microscopy.

Analysis of the E-pH diagrams shows [2], that oxidation of CuS to elemental sulphur may be expected only at very low pH values and/or at very low concentration of free copper (II) ions in the solution. These observations coupled with current efficiency studies allowed to establish that pH increase causes the change of predominant anodic reaction from CuS $\longrightarrow$ Cu^{2+} + S + $2e^-$ at pH=0,

to: $CuS + 4H_2O \longrightarrow Cu^{2+} + SO_4^{2-} + 8H^+ + 8e^-$ at pH=2,

while the increase of chloride concentration influences the anodic dissolution as a result of the chemical reaction: $CuS + 2Cl^- \longrightarrow CuCl_2^- + S$. The intercombinations of those reactions is responsible for anodic behaviour of copper (II) sulphide.

CONCLUSIONS

It is possible to obtain copper solid-state ISE with membranes containing pure copper (II) sulphide. However, this goal is a black art because avoiding membrane desintegration needs highest purity CuS but with increasing purity of CuS the chemical diffusion of copper in this compound is lowered.

The chlorides strongly influence the electrode processes participating in the irreversible anodic reactions leading to the formation of elemental sulphur.

REFERENCES

1. E.Ghali, B.Dandaponi and A.Lewenstam, Surface Technology, in print.
2. E.Ghali, B.Dandaponi and A.Lewenstam, J.Electrochem.Soc., in print.
3. A.Lewenstam, A.Hulanicki and E.Ghali, - in preparation.

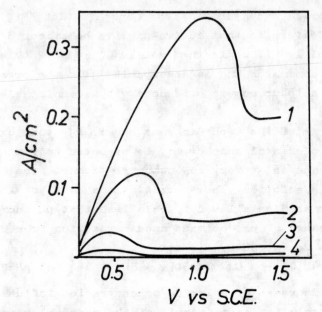

Fig. 1. Potentiodynamic curves in 1 mol/1 HCl for Ventron
CuS. Scan rate: 1 - 100 mV/sec, 2 - 10 mV/sec, 3 -
1 mV/sec, 4 - 0.1 mV/sec.

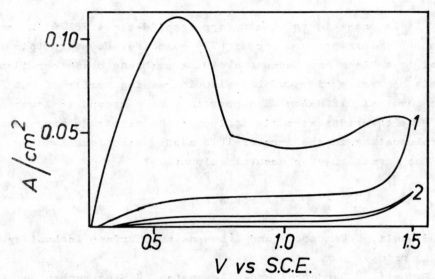

Fig. 2. Cyclic voltammograms for Ventron CuS at scan rate
10 mV/sec. 1 - first, 2 - tenth cycle.

238

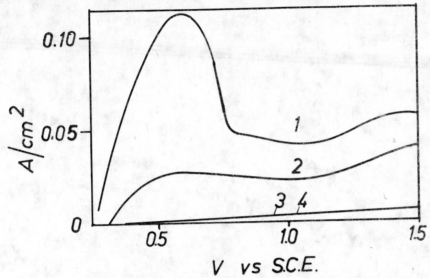

Fig. 3. Potentiodynamic curves for CuS. 1 – Ventron CuS in 1 mol/l HCl, 2 – home-made CuS in 1 mol/l HCl, 3,4 – Ventron and home-made CuS in 10^{-2} mol/l HCl.

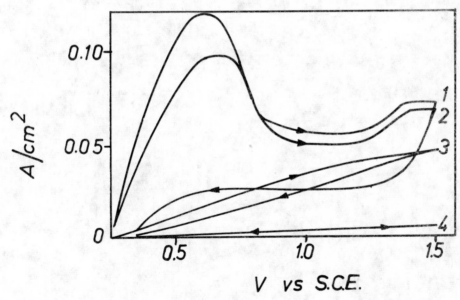

Fig. 4. Influence of hydrogen and chloride ions concentration on the voltammetric behaviour of Ventron CuS electrode at 10 mV/sec. 1 – 1 mol/l HCl, 2 – 10^{-2} mol/l KCl, 3 – 10^{-2} mol/l HCl + 10^{-1} mol/l KCl, 4 – 10^{-2} mol/l HCl.

A GLUCOSE-SELECTIVE SENSOR FOR THE DETERMINATION OF GLUCOSE IN BLOOD AND URINE

J. HAVAS,* E. PORJESZ,* G. NAGY** AND E. PUNGOR**

*Department for Research and Development, Radelkis Electroanalytical Instruments, Budapest, Hungary
**Institute for General and Analytical Chemistry, Technical University, Budapest, Hungary

ABSTRACT

A new type of glucose-selective sensor operating with immobilized glucose oxidase enzyme was developed and its preparation, electrochemical behaviour and applicability were studied.

The electrode showed a long life-time and a short response time.

The glucose response of the sensor was found to be independent of the original oxygen concentration of the supporting solution in the concentration range of analytical importance.

A method was worked out for measuring the glucose content of biological fluids. The accuracy and applicability of the technique were extensively studied on a large number of artifical and biological samples.

INTRODUCTION

The rapid and selective determination of glucose in the sample solutions of small volume is of especially great importance in the clinical laboratory practice. The problem is most frequently solved by the use of a glucose-selective sensor /1-15/ and the amperometric detection is in the focus of interest because of rapidness, simplicity and low specific cost of this technique.

The use of molecule-selective sensors is limited by the fact that the preparation of the sensor having a long life-time

241

and well reproducible reaction layer has been seldom success-
full up till now /16/. Recently a short paper was published
about a new type of glucose-selective sensors, which can be
prepared easily and has a long life-time /17/. The study of
the electrochemical behaviour and applicability of this glucose
sensor is summarized in this work.

The glucose-selective sensor is a sensitised voltammetric
measuring cell covered with an active membrane containing
immobilized glucose oxidase enzyme.

The operation of the sensor is based on the oxidation of
glucose by the glucose oxidase enzyme. The reaction

$$\text{glucose} + O_2 + H_2O \xrightarrow[\text{E.C. 1.1.3.4.}]{\text{glucose oxidase}} \text{gluconic acid} + H_2O_2$$

can be followed by the measurement of the decrease in the
oxygen concentration by means of an amperometric electrode.

If the chemical reaction and the mass transports attain
a steady state in the reaction layer - in a given substrate
concentration range - a well defined relationship exists
between the decrease of the amperometric current and the
glucose concentration of the sample.

EXPERIMENTAL

The sensor /Radelkis type OP-Gl-7113/ was an amperometric
electrode, essentially identical with the Clark-type electrode.
The working electrode was made of 4 platinum wires soldered
in a glass tube. The diameter of the individual wires was 0.3
mm and the filling solution was a RADELKIS oxygen sensor
electrolyte, type "O-107".

A protein-based membrane /pig small intestine/, with the
immobilized glucose oxidase enzyme on its surface, was used
as the active membrane. The reaction layer containing the immobilized enzyme was prepared in the following way: 10 mg
bovinalbumen and 40 mg glucose oxidase enzyme were dissolved
in 0.5 m^3 of phosphate buffer of pH=7.0 at room temperature.
After the complete dissolution /about 2 minutes/ 25 µl of

242

glutaric aldehyde was added, the mixture was rapidly and thoroughly homogenized and a 20 μl portion of the solution was dropped onto and spread on the protein membrane surface. The membrane was kept at room temperature for 1 to 2 hours. During this time the reaction layer was formed. The thickness of the active membrane was about 10 μ. The active membrane with the polypropylene membrane of the Clark cell assembled in a "sandwich" structure together with the "O" rings were placed into the screwed cap of the sensor. Hereupon the cup was attached to the electrode body.

The measurements were carried out in a glass cell containing a magnetic stirring rod and equipped with a thermostated jacket.

In order to be able to use the glucose sensor in connection with a conventional pH-meter, a Glucose Adapter /RADELKIS type OP-960/ was developed. It supplies a polarization voltage of -0.65 V for the working electrode and converts the cell current to a voltage signal. The latter was measured with a Laboratory digital pH-meter /RADELKIS type OP-211/1/.

The sensor was stored in a phosphate buffer of pH = 7.0 at room temperature, if frequently used, in the opposite case it was disassembled and kept dry at room temperature.

RESULTS AND DISCUSSION

The response time of the sensor was studied by the immersion measuring technique. The glucose sensor was immersed in an intensively stirred, thermostated /310°K/ phosphate buffer of pH = 7.0 and the voltammetric current-time curve was recorded continuously. Hereupon 0.1 cm^3 portion of the glucose solution of known concentration was added into the buffer instantaneously. As a result of the addition of glucose solution a change could be observed in the voltammetric current. After a short time the current signal stabilized.

The time period between the beginning of the current change and the instant to 90 per cent of the total current change was defined as the response time of the sensor.

The response time was found to be 40 to 55 seconds depending on the glucose concentration /4 to 10 mmol/dm^3/ in the sample.

Comparing the measured response time values with the literature data /11, 12, 14/, it can be concluded that the measured values are smaller than those observed with other enzyme electrodes.

The life-time of the sensor were studied with 15 sensors prepared by identical procedures. The values /Δi_c/ obtained after the change of the glucose concentration from 0 to 0.15 mmol/dm^3 in the solution, were measured once a month.

The sensors had a long life-time: even after 240 days the activity of the active membrane was as high as 60 to 80 per cent of the initial value. Despite of this decrease in the activity the sensors could still be well used for the glucose determinations.

The life-time of the sensors can probably be increased by storing them in refrigerator.

The dependence of the glucose response of the sensor on the original oxygen concentration of the supporting solution /phosphate buffer of pH = 7.0/ was also examined.

Nitrogen and oxygen gas mixtures of different compositions were prepared by varying the flow rates of the gases effusing from gas cylinders. The gas mixture saturated with water vapour was bubbled through 20 cm^3 of phosphate buffer of pH = 7.0 placed in the measuring cell. The temperature of the buffer was kept at 298°K and stirred at constant rate. The oxygen concentration of the supporting solution was determined with pO$_2$ sensor of RADELKIS. Hereupon 50 μl portion of glucose solution of 0.1 mol/dm^3 concentration was added to the buffer and the change in the current was measured. The sensor signal values plotted against the original oxygen concentration of the supporting solution are shown in Figure 1. It is apparent from the figure that the change in the voltammetric current of the sensor caused by an identical amount of glucose is - over a wide range - practically independent of the original oxygen concentration of the supporting solution.

Accordingly, the measurements can be carried out in a

supporting solution in contact with the atmosphere.

The response of the sensor was plotted against glucose concentration /Figure 2/. It is apparent from the figure that the concentration-dependence of the signal is linear in the 2 to 20 mmol/dm^3 range.

The amperometric glucose measurements were carried out in the concentration range corresponding to linear response. The measurements were made by the standard addition technique in the following way: first 100 μl of the glucose solution of known concentration was added to the supporting solution and the decrease in the current was measured. Hereupon 100 μl portion of the sample was added to the same solution. Measuring the change in the current again the glucose concentration in the sample was calculated from the ratio of the changes in current.

In order to characterize the reproducibility of this series of glucose determinations were carried out with standard solutions of different glucose concentrations. The reproducibility was found to be dependent on the glucose concentration in the sample. The results are summarized in Table 1. It is apparent that the reproducibility in the concentration range of analytical importance is $\pm$ 5 per cent.

In order to characterize the accuracy of the measuring technique several measurements were carried out with commercially available freeze-dried plasma samples. The glucose concentration in these samples were specified by the manufacturer. The results of these glucose determinations are summarized in Table 2. It is apparent that the results agree with the nominal values within the experimental error.

The applicability of the method was extensively studied on a large number of biological samples. The comparative measurements were carrier out with 26 heparinized blood, 58 different plasma and 19 urine samples. The equation of the regression line and the corresponding correlation coefficients /f/ calculated from the results are given below:

$$y = 0.96 \, x - 0.36; \quad f = 0.989 \qquad \text{for blood samples}$$
$$y = 0.96 \, x + 0.32; \quad f = 0.995 \qquad \text{for plasma samples}$$
$$y = 1.02 \, x - 0.01; \quad f = 0.997 \qquad \text{for urine samples}$$

The results of the comparative measurements obtained with plasma samples are shown in Figure 3. The values determined amperometrically are plotted againts the glucose concentration values measured by spectrophotometric technique.

These measurements demonstrate that the electrochemical technique is by no means inferior to the optical one. The reproducibility of the latter may be somewhat greater; however, the high speed of the electrochemical measurements - which e.g. in intensive care units can be of great importance - furthermore the possibility for continuous following of changes in glucose concentration - e.g. in fermentation - appear to tilt the balance in favour of the electrometric technique.

REFERENCES

1. J.Sós, Laboratóriumi diagnosztika. Medicina Press, Budapest, 1974
2. M.M.Fishman and H.F.Schiff, Anal.Chem. 46, 367 R /1974/
3. Beckman Bulletin G-308
4. L.D. Bowers and P.W.Carr, Anal.Chem. 48, 545 A /1976/
5. B.Danielsson, K.Gadd, B.Mattiasson and K.Mosban, Clin. Chim.Acta 81, 163-175 /1977/
6. S.Updike and G.P.Hicks, Nature 214, 986 /1967/
7. W.D.Doig and A.Körösi, Anal.Chem. 42, 118 /1970/
8. L.C.Clark, Proc.Internat.Union.Physiol.Sci. 9, /1971/
9. G.G.Guilbault and G.J.Lubrano, Anal.Chim.Acta 60, 254 /1972/
10. G.G.Guilbault and G.J.Lubrano;Anal.Chim.Acta 64, 439 /1973/
11. G.Nagy, von L.W.Storp and G.G.Guilbault, Anal.Chim.Acta 6, 443 /1973/
12. S.P.Bessman and R.D.Schultz, Ion-Selective Micro-Electrodes. Plenum Press, New York, London, 1974
13. L.D.Mell and J.T.Maloy, Anal.Chem. 47, 299 /1975/
14. S.C.Martiny and O.J.Jensen, Ion and Enzyme Electrodes in Biology and Medicine. Urban and Schwarzenberg Press, München, Berlin, Wien, 1976

15. J.Havas, Ion and Molecule-Selective Electrodes in Biological
    Fluids. Akadémiai Kiadó, Budapest, 1980
16. G.Nagy and E.Pungor, Hung.Sci.Instr. 32, 1 /1975/
17. J.Havas, Magy.Kém.Folyóirat 85, 329 /1979/

Table 1. Reproducibility of the measuring method in the case
          of different sample concentrations

Mean value		Reproducibility		
$mmol/dm^3$	mg%	$mmol/dm^3$	mg%	%
1.7	30	± 0.20	± 3.6	± 12.0
3.3	60	± 0.20	± 3.6	± 6.0
5.0	90	± 0.25	± 4.5	± 5.0
6.7	120	± 0.30	± 5.4	± 4.5
10.0	180	± 0.35	± 6.3	± 3.5
13.3	240	± 0.50	± 9.0	± 3.8
20.0	360	± 0.80	±14.4	± 4.0

Table 2. Summary of results of the determination of the glucose
          concentration of commercial freeze-dried plasma samples

Plasma product	Nominal value $mmol/dm^3$	Measured value $mmol/dm^3$
Precinorm	5.2 ... 6.4	6.4
Bio Merieoux	14.1 ... 15.3	15.1
DADE	13.1 ... 17.1	15.5
HYLAND	9.3 ... 10.7	10.2

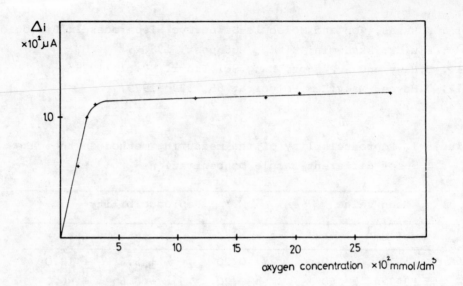

Figure 1. The dependence of the glucose response of the sensor on the original oxygen concentration of the supporting solution

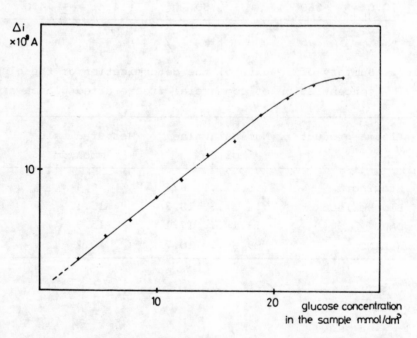

Figure 2. The dependence of the response of the sensor on the glucose concentration in the sample

248

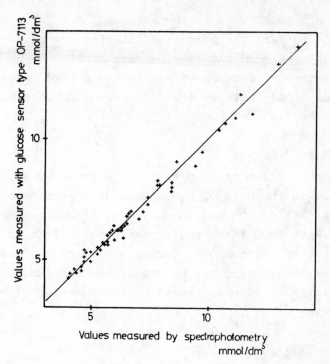

Figure 3. The regression line of the comparative measurements
obtained with plasma samples

QUESTIONS AND COMMENTS

Participants of the discussion: B.Jänchen, E.Porjesz, G.Nagy

Question:
The oxygen signal was found to be unsuitable for the measurement of blood, because the latter has an oxygen consuming
reaction. How can your results be interpreted from a theoretical point of view ? What kind of amperometric sensor did
ensure such a good long-term stability ?

Answer:
The oxygen consuption of blood itself is a relatively slow
process. It just can cause a relatively small change in the
overall oxygen concentration of the sample. Furtheron, the

249

blood sample is diluted in the measuring cell with air-
-saturated buffer. Therefore the interference caused by the
effect you mentioned can only be very small.

As a matter of fact, many publications have appeared on
successful blood analyses based on enzymatic reaction and
the Clark oxygen electrode.

Comment:

A very stable basic sensing element is required for an enzyme
electrode. In the case of H_2O_2 sensor film formation occurs
at the surface of the basic sensing electrode, which is not
the case with the Clark-type oxygen sensor. That is the reason
why the latter was employed for constructing a glucose
electrode. Considering the volume of the samples, about 1%
systematic error is only caused by the reaction of haemoglobine
with oxygen.

ANION INTERFERENCES OF CALCIUM SELECTIVE ELECTRODE

A. HULANICKI, M. TROJANOWICZ AND Z. AUGUSTOWSKA

Department of Chemistry, University of Warsaw, Warsaw, Poland

Anion interferences of liquid-state cation selective electrodes received much attention in the case of membrane containing neutral carriers. Less probable seems to be a similar effect for membranes with negatively charged sites, however such interferences for the calcium selective electrodes based on phosphoric acid esters has been mentioned in the literature [1-3]. In this paper a more detailed study of effects of membrane matrix and role of mediator is presented.

Our measurements with PVC membrane confirmed the opinion that this material is a completely inert matrix when both calcium chloride or calcium perchlorate solutions were investigated. Measurements with the cellulose Millipore membranes gave conclusions that for low calcium salt concentrations the cationic functioning is observed and anionic response for higher salt levels independently whether calcium chloride or perchlorate were used /Fig. 1/.

The cationic function of Millipore membranes saturated with di/n-octyl/phenylphosphonate /DOPP/, towards calcium is much better developed being however strongly influenced by the salt anion /Fig. 2/. For nitrate, but especially for thiocyanate and perchlorate the anion interference is pronounced and limits the upper concentration range. In the case of two latter mentioned anions a maximum at the response curve at approximately pCa 3.0 was observed. Less significant interferences of anions were observed for the same membrane matrix saturated with decanol /Fig. 2/. This suggests

251

that the mediator plays an important role in anion interferences and that DOPP containing membranes are influenced by
penetration of lipophilic anions into the membrane phase. The
PVC membrane with DOPP behaves differently /Fig. 3/. While in
chloride, bromide and to a smaller degree in nitrate solutions the electrode response is nearly Nernstian up to pCa 3,
in calcium perchlorate media the electrode characteristics
is irregular and mainly over-Nernstian in the range of concentrations above pCa 3.

The anion interferences of the complete electrode system
of the Orion model 92-20 calcium electrode do not differ significantly from those observed in the absence of the ion-exchanger in the membrane phase. Similarly behaved the calcium
selective PVC electrode containing a mediator and an ion-exchanger although larger anion interferences in the presence
of chloride and nitrate were observed /Fig. 4/. In comparison
to PVC membrane with DOPP, irregularities of the response
function disappeared and a pronounced anion-function develops.
Experiments were also performed with the exchanger diluted
1:1 and 1:5 with DOPP. Dilution of the exchanger in the membrane shifts the maximum on the electrode response curve towards higher pCa values, and at dilution 1:5 again a cationic
function is restored at pCa 2. Such effect seems to be an intermediate step to the behaviour of PVC electrode devoid of
ion-exchanger /Fig. 3/. This conclusion is in agreement with
suggestions of Buck et al. [4] who describe the position of
the maximum on the electrode response curve as a function of
the concentration of sites in the membrane.

The explanation of the pronounced role of mediator in the
anion interferences of the calcium selective electrode is based on interaction of mediator with calcium ions by formation
of a positively charged complex in the membrane, having the
formula:

$$\left[(RO)_2P(R')O \longrightarrow Ca\right]^{2+}$$

which acts as an anion exchanger and in the presence of lipophilic anions shows the anionic function. In the higher con-

centration range of such anions, partial penetration of membrane phase by anions occurs. Therefore, the type of mediator is of importance and significant effects in the presence of DOPP are much smaller than for decanol containing membranes. The ion-exchanger in the porous membrane has a minor effect on the anion function, but is vital for the electrode selectivity. The PVC membranes behave differently. In the absence of ion-exchanger, with DOPP as mediator they show over-Nernstian function at high calcium concentration. The increase of active site concentration favours the elimination of both irregularities and anion interferences.

REFERENCES

1. J.W.Ross, Science, 156, 1378 /1967/
2. G.A.Rechnitz and Z.Folin, Anal.Chem., 40, 696 /1968/
3. J.Růžička, E.H.Hansen and J.Chr.Tjell, Anal.Chim.Acta, 67, 155 /1973/
4. R.P.Buck, F.S.Stover and D.E.Mathis, J.Electroanal.Chem., 82, 345 /1977/.

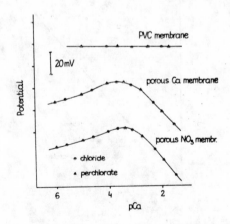

Fig. 1. Potential response of electrode matrix material in calcium chloride and perchlorate media

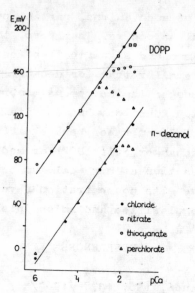

Fig. 2. Potential response of electrodes with porous membrane containing mediator in solutions of calcium salts

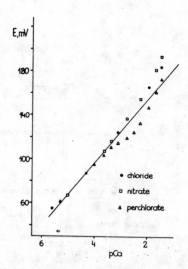

Fig. 3. Potential response of PVC membrane electrode containing di/n-octyl/phenylphosphonate in solutions of calcium salts

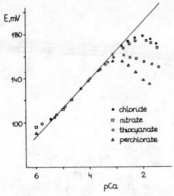

Fig. 4. Potential response of PVC electrode with calcium exchanger /ORION 92-20-02/ in solutions of calcium salts

QUESTIONS AND COMMENTS

Participants of the discussion: R.P.Buck, M.Trojanovicz

Question:
Is it possible to interpret the slopes of the curves of the anion effect ? Can you measure the transference numbers or the number of sites or collect any data to fit any model of the sequence as a result of your work ?

Answer:
No, unfortunately not. It was only potentiometric measurement, and we have found in the initial stage, during the routine use of a PVC calcium electrode a curvature at higher concentrations, so it was decided to study this effect further. A guess might be that this is a result of bonding between calcium ions and phosphate in the membrane phase, but we did not intend to go further with the interpretation.

ABOUT THE DETERMINATIONS OF TETRAETHYL-LEAD CONTENT OF GASOLINES USING ION-SELECTIVE ELECTRODES

M. S. JOVANOVIC,* M. DJIKANOVIC,** B. D. VUCUROVIC AND Z. ABRAMOVIC*

*Institute for Analytical Chemistry, Faculty of Technology and Metallurgy,
University of Beograd, Yugoslavia
**Institute for Inorganic and Analytical Chemistry, Faculty of Metallurgy,
Titograd, Yugoslovia

ABSTRACT

The internationally accepted standard methods for the
determination of tetraethyl-lead-content in commercial gasolines i.e. the gravimetric determinations according to the
ASTM as well as the Scott's Standard Methods, are substituted
in these investigations by potentiometric procedures. Ion-
-selective electrodes of all-solid-state types with either
Ag_2S or PbS/Ag_2S sensors were used, enabling both usual
titrimetric determination and direct measurement of the e.m.f.
of ion-selective electrode/examined solution/SCE couple. In
the first case, lead ions in the solution were titrated with
a standard solution of potassium chromate while in the second
one, the obtained values were plotted on a calibration graph.
The results obtained by titrimetric potentiometric method are
in good agreement with those of standard procedure, while
direct potentiometry yields some higher values.

INTRODUCTION

The content of tetraethyl-lead /TEL/ is an important factor
influencing the Octane Number /ON/, the main characteristic
of all gasolines for internal-combustion engines. The Octane
Number itself expresses the resistance against self-detonation
of the fuel when engine is running. To improve the quality of

257

gasolines obtained by distillation, anti-detonating agents, chiefly tetraethyl-lead, are added. So, there are some quantities of lead in all the exhausted gases of such engines, but because of the danger to the environment, the TEL content of gasolines is limited by low.

Edgar and Calingaert /1/ were among the first who developed a method for the determination of TEL content in gasolines and their technique was accepted by ASTM /2/ and Scott's Standard Methods /3/ as one of standard methods. Besides this gravimetric method, there are also some others described in the literature /4,5,6/ but it seems that Edgar and Calingaert's method is mainly applied. For the reason mentioned, this method consisting of 1. extraction of lead from gasolines in form of $PbBr_2$ with carbon-tetrachloride solution of bromine, 2. conversion of obtained precipitate into solution of $Pb/NO_3/_2$ with concentrated nitric acid, 3. dilution of the obtained solution and adjusting of the pH value followed by addition of potassium dichromate in excess and 4. drying and weighing of the obtained lead chromate, was accepted as the reference one for these investigations.

Our intention was to substitute long-lasting gravimetry with potentiometric titrimetry, applying two first phases of Edgar and Calingaert's method and using ion-selective electrodes. Our final goal was to substitute even titrimetry with direct potentiometric measurement of the e.m.f. estabilished by two electrodes in the examined solution. For this reason, three all-solid-state ion-selective electrodes were used: a commercial Ag_2S sensed one /giving a response also if Pb^{2+}-ion concentration changes during the titration with an anion forming a precipitate with Ag^+ ions, too/, a commercial lead--selective electrode of universal Ruzicka type, and a self-made lead-selective one also of Ruzicka type.

EXPERIMENTAL

Instrumentation

A Radiometer M 62 Standard pH-meter was used for measuring e.m.f. established by ion-selective electrode/examined lead-

258

-ion solution/saturated calomel electrode couple.

A Metrohm E 485 piston-burette equipped with a magnetic stirrer was applied for titrimetric procedures.

All-solid-state ion-selective electrodes used were the following:
- Radiometer F 1212 S selective electrode with a crystal of Ag_2S as the sensor,
- Radiometer F 3012 universal, Ruzicka type Selectrode,
- self-made universal, Ruzicka type selective electrode sensed with a mixture of PbS/Ag_2S.

Preparation of self-made electrode

The carrier of the sensor was made by mixing of graphite and Teflon powders, both of them passing a sieve of 0,20 mm. The weight-ratio of graphite against Teflon was from 30 : 70 to 5 : 95, however, that one of 10 : 90 showed to be the best. The mixtures of these powders, homogenized as well as possible, were then hydraulically pressed by 2 tons/cm^2 giving finally a cylindrical body 5,9 mm of diameter and 32 mm of its length. Ohmic resistance of such a sensor-carrier was found to be 40 ohms, and the body itself showed good mechanical strength. The body was put into a Teflon tube of 5.9 mm inner diameter and 10 cm of length. The electrical connection between the body of the electrode and the cable of the high-impedance input of the pH-meter, was made by means of a graphite brush of a small electro-motor, its copper spring was soldered to the conducting wire.

Active material to be pressed onto the exposed surface of the electrode body, was prepared according to the description of Ruzicka and coworker /7/. Here, the mol ratio of lead sulphide of 4 : 1 showed the most similar behaviour to that of commercial F 3012 electrode. This was achieved mixing 20 cm^3 of 0.2 mol/dm^3 of lead nitrate solution with the same volume of 0.1 mol/dm^3 of silver nitrate solution, followed by the addition of 60 cm^3 of 0.1 mol/dm^3 sodium sulphide. After half an hour waiting, the obtained precipitate was filtered through a crucible, thoroughly washed with distilled water and dried at room temperature.

Only very small quantities /a few mgs/ of the sensor obtained in this way, should be rubbed into the exposed surface of the electrode body by help of a rounded glass rod.

Chemicals

All the reagents used were of analytical grade purity.

As titrant served a 0.01 mol/dm^3 solution of potassium chromate, the same substance which is used for precipitation of lead chromate in gravimetric procedure.

For construction of calibration graph solutions of lead perchlorate from 10^{-1} to 10^{-5} mol/dm^3 were used, in order to avoid complexation of lead by nitrate ions.

For the extraction of lead from gasolines, a 10% solution of bromine in carbon tetrachloride was applied.

Two different sorts of gasolines, one with lower the other with higher Octane Number, got from fuel stations, were taken for determinations of TEL content.

PROCEDURE

Previous investigations

The electrochemical behaviour of our, self-made ion-selective electrodes were compared with the behaviour of commercial F 3012 Selectrode in the same solution of lead perchlorate. Going from weight ratio of graphite to Teflon from 30 : 70 /G30T70/ to 5 : 95 /G5T95/, the slope of a stright line /being curved slightly only below 10^{-4} molar concentrations/ representing the dependence of the ion-selective electrode potential on lead-ions concentration, expresses all the better tangent. In all cases, the response time was not longer than 20 seconds, however, the potentials of G5T95 self-made electrode were rather unstable. Potential-dependence line of G10T90 electrode compared with that of commercial F 3012 one, is shown on Fig.1.

The self-made G10T90 /graphite : Teflon = 10 : 90/ electrode bearing a PbS/Ag$_2$S sensor in mol ratio of 4 : 1, compared with F 3012 electrode, showed also applicability for

titrations of pure lead nitrate solutions with standard
solution of potassium chromate, as shown on Fig. 2.

Titration curves, very similar in shape were also obtained
if Ag_2S sensed, commercial F 1212 S electrode was used under
same conditions.

Determinations of TEL content of gasolines

Regardless if gravimetric, titrimetric or direct potential-
-reading procedures were applied, the extraction of lead from
gasolines was the same. Samples of 100 cm^3 of gasolines should
be treated with so much carbon tetrachloride /bromine solution
as to obtain a dark brown colour /15-25 cm^3/. After 15 minutes,
the obtained precipitate of $PbBr_2$ should be filtered through
a crucible, washed with petroleum ether and dissolved in
concentrated nitric acid. In order to eliminate the excess
of bromine, solution of lead nitrate should be twice evaporated
to dryness. After the addition of about 50 cm^3 of water and
adjusting the pH value to 6 with ammonia, the solution of lead
is ready for further treatment.

If gravimetric determination is intended, a few cm^3 of
2 mol/dm^3 acetic acid should be added previously, followed by
the addition of 10 cm^3 of 5% potassium chromate solution.
After half an hour waiting, obtained lead chromate is filtered
through a crucible, washed and dried to constant weight.
Weight of lead chromate represents the quantity of tetraethyl
lead in 100 cm^3 of gasoline, because mol-masses of both the
substances are the same.

In the case of potentiometric titration, the solution of
lead acetate obtained as described in the foregoing paragraph,
should be titrated with standard 0.01 mol/dm^3 solution of
potassium chromate applying any of the mentioned ion-selective
electrodes coupled with saturated calomel one.

Finally, if direct potential-reading procedure is intended,
solution of lead nitrate adjusted to pH=6 with ammonia /with-
out the addition of acetic acid/, should be diluted with
distilled water just to 100 cm^3. Here, potential-readings can
be made using PbS/Ag_2S sensed electrodes only. So, any of

F 3012 or GlOT90 coupled with SCE should be immersed into the solution and the read value of potential drawn onto the calibration graph.

RESULTS AND DISCUSSION

As can be seen /Fig. 1/ electrochemical behaviour of the self-made GlOT90 ion-selective electrode is practically the same as that of commercial F 3012 one. The first electrode shows even a less curved potential-dependence on lead ion concentrations below 10^{-4} molar solution. On the other hand, there is a slight sub-Nernstian slope in both the cases. There is a similar situation if electrodes are used for potentiometric titrations /Fig. 2/.

In Table 1 are compared results of the determinations of two examined gasolines using the three techniques: gravimetric, potentiometric and direct potential-readings. Here, it must be underlined that the application of self-made GlOT90 electrode yields no results, if direct potential-reading is made. It appeared namely to be just impossible removing last small quantities of organic residue from lead nitrate even after repeated evaporation to dryness. After dilution with water small dark-brown spots of organic materials disperse into the solution and their presence is possible interfering. It is also possible that the same leads to some higher results even if F 3012 electrode is used for direct potential-reading. Finally, it should be stressed that the results are remarkable higher than the expected values in the case of gasoline with higher Octane Number. Because the lead concentration in the solution to be determined is below 10^{-3} molar, it is hard to prescribe the deviations of results, to some complexations of lead by nitrate ions. On the other hand, the fact is that the quantity of organic residue in such a case is greater than if gasoline with lower Octane Number is examined. This, perhaps, can be the explanation for the deviations observed.

262

REFERENCES

1. R. Edgar and M. Calingaert, Ind.Eng.Chem., Anal.Ed., $\underline{1}$, 221 /1929/
2. ASTM, 1961, Part 7, p.270
3. W.W.Scott, Standard Methods of Chemical Analysis, V.Nostrand, London, Vol.II, p. 270, 1955
4. ASTM, 1961, Part 7, p. 679
5. A.Grünwald, Erdöl u. Kohle $\underline{6}$, 550 /1953/
6. M.S.Jovanović, J.Tomić, Z.Masis and M.Dragojević, Chem.anal. /Warszawa/ $\underline{11}$, 478 /1966/
7. E.H.Hansen and J.Ruzicka, Anal.Chim.Acta, $\underline{72}$, 365 /1974/

Table 1   Results of TEL-content determinations in gasolines using different procedures

Method of determination	Number of analysis	Found TEL in g/dm³ /mean value/	
		Gasoline A	Gasoline B
gravimetric	10	0,5640	0,7941
potentiometric titration			
- F 1212 S	10	0,5666	0,8169
- F 3012	10	0,5658	0,8150
- G10T90	10	0,5663	0,8172
direct potential-reading			
- F 3012	10	0,6023	0,8678

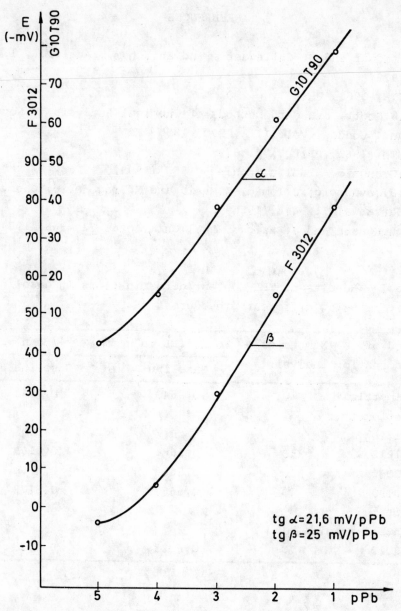

Figure 1. Calibration graphs of lead-selective electrodes
in lead perchlorate solutions

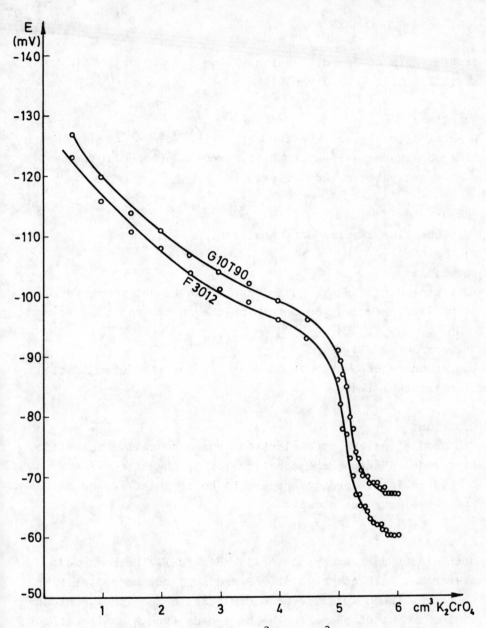

Figure 2. Titrations curves of 10^{-3} mol/dm^3 lead nitrate solution determined with 10^{-2} mol/dm^3 potassium chromate solution using lead-selective electrodes

QUESTIONS AND COMMENTS

Participants of the discussion: E.Pungor, J.D.R.Thomas,
E.Vasilikiotis, M.S.Jovanovic

Question:
Lead is often present in the air in the form of oxides. How
did you determine these species ? How did you collect the
sample and perform the analysis ?

Answer:
Only gasoline samples were analyzed.

Question:
Your procedure appears to be unsuitable to decompose tetraethyl lead. How did you determine the latter compound ?

Answer:
The standard chromate method was used for the determination
of tetraethyl lead.

Question:
You used a Ruzicka type electrode and a so called home-made
electrode. What was the constitution of the latter ? A
considerable difference appears to be in the E^O value of
the two electrodes shown.

Answer:
Both electrodes were practically identical. The home-made
electrodes were composed of PbS and Ag_2S compressed at high
pressure. The only difference could be in the inner contacts
of the two electrodes. In the home-made type a small carbon
brush contact was inserted in the electrode.

Question:
Can you use your method to determine the lead content of air ?

Answer:
Our method was not applied for air pollution analysis yet.

DIFFUSION PHENOMENA: THEIR EFFECT ON THE CHARACTERISTICS OF ION-SELECTIVE ELECTRODES

W. E. MORF

Swiss Federal Institute of Technology, Department of Organic Chemistry,
CH-8092 Zürich, Switzerland

ABSTRACT

The influence of surface-film diffusion processes on the dynamic response and the selectivity behavior of ion-selective electrodes is discussed. A diffusion model of liquid-membrane electrodes (a generalized version of the Jyo-Ishibashi theory) is presented that accounts for the observable phenomena. The theory encompasses earlier approaches to the apparent selectivities or to the transient responses of ion sensors, and it fully agrees with experimental findings.

INTRODUCTION

The response mechanism of ion-selective electrodes, gas sensors, and enzyme electrodes involves diffusion of ions or reactants from the bulk sample solution to the ion-sensing electrode surface. As a consequence, the activities measured by the electrochemical cell must not necessarily correspond to the intrinsic sample activities. This may give rise to time-dependent response phenomena (response delay, transient response) and to variations in the apparent selectivity coefficients. Some fundamental diffusion phenomena and their ·consequences are discussed and summarized in the following.

DYNAMIC RESPONSE TO THE PRIMARY ION

In the absence of interferents, the idealized response of ion-selective electrodes to primary ions $\underline{i}$ is given by the Nernst equation:

$$E = E_i^O + s \cdot \log a_i' \tag{1}$$

where E is the e.m.f., E_i^O the standard potential of the cell, and s the slope of the response function. The measured activity a_i' refers to the boundary zone of sample solution contacting the ion-sensing membrane surface. This value deviates from the bulk sample activity a_i as long as diffusion processes are not terminated. Thus, following a step change $a_i^o \rightarrow a_i$ in the bathing solution at the time $t = 0$, the sensed activity generally becomes

$$a_i'(t) = a_i - (a_i - a_i^o) \cdot f(t) \tag{2}$$

where $f(t)$ is a characteristic function decreasing from 1 for $t = 0$ to 0 for $t \rightarrow \infty$.

If diffusion is restricted to the unstirred layer of sample solution adhering to the ion-sensing membrane, $f(t)$ assumes the form [1-5]:

$$f(t) = \frac{4}{\pi} \sum_{n=0}^{\infty} \frac{(-1)^n}{2n+1} \exp\left[-\frac{(2n+1)^2 t}{\tau'}\right]$$

$$\approx e^{-t/\tau'} \quad (\text{for } t > \tau') \tag{3}$$

where the time constant τ' depends on the thickness δ' of the external diffusion layer and on the ionic diffusion coefficient D':

$$\tau' = \frac{4}{\pi^2} \frac{\delta'^2}{D'} \approx \frac{\delta'^2}{2 D'} \tag{4}$$

A completely different relation for $f(t)$ obtains if diffusion into the ion-selective membrane is rate-controlling [1-4]:

$$f(t) \cong \frac{1}{\sqrt{t/\tau} + 1} \tag{5}$$

The new time constant τ includes contributions from the internal diffusion coefficient D and a distribution parameter K characteristic of the membrane:

$$\tau = \frac{D K^2 \delta'^2}{\pi D'^2} \tag{6}$$

A square-root time dependence of the type (5) was established for valinomycin-based PVC membrane electrodes which, under optimal conditions, show τ-values of only a few milliseconds [3,4]. In contrast, an exponential time response was found for a valinomycin/silicone rubber membrane electrode ($\tau' \approx 0.2$ s) [4] and for a calcium-selective liquid ion-exchanger electrode [6] ($\tau' \approx 1$ s [1,2]). Thus, for general cases, the dynamic behavior of diffusion-type (liquid) membranes may be characterized by the following approximation:

$$f(t) \approx \frac{1}{\sqrt{t/\tau}} + e^{-t/\tau'} \qquad (7)$$

which is valid for $t \gg \tau$ and $t > \tau'$ [2].

SELECTIVITY BEHAVIOR AND DYNAMIC RESPONSE IN THE PRESENCE OF INTERFERING IONS

The response of an ion-selective electrode to a mixed solution of two ions i and j, both having the same charge z, is usually given by the Nicolsky equation:

$$E = E_i^o + s \cdot \log [a_i' + K_{ij} a_j'] \qquad (8a)$$

Corresponding theoretical expressions were derived for the following types of ion sensors [1]: (a) anion-selective solid-state membrane electrodes, (b) electrodes based on dissociated or non-specific liquid ion-exchangers, (c) divalent-ion-selective electrodes based on liquid ion-exchangers, (d) neutral carrier membrane electrodes, and (e) idealized glass electrodes. The theoretical selectivity factors K_{ij} are related to the ratios of solubility products (a), ionic distribution coefficients (b), and stability constants of ion/ligand complexes (c,d), or to the ion-exchange equilibrium constant (e) [1].

In practice, however, the potentiometric selectivity coefficients are defined in terms of bulk sample activities, i. e.

$$E = E_i^o + s \cdot \log [a_i + K_{ij}^{Pot}(app.) \, a_j] \qquad (8b)$$

Diffusion-controlled deviations between the intrinsic sample activities (a_i, a_j) and the values (a_i', a_j') measured at the electrode surface may give rise to characteristic variations in the apparent selectivities. Thus it is often observed that the values of $K_{ij}^{Pot}(app.)$ for a given measuring system tend to converge on unity with decreasing activities of the sample solution. At very low activities, the electrode seems to measure primary ions eluted from the membrane rather than the ions coming from the sample solution.

269

An interpretation of the apparent selectivity behavior of liquid ion-exchange membrane electrodes has been given informally by Hulanicki and Lewandowski [7] and more recently in a theory by Jyo and Ishibashi [8, 9]. In the latter treatment, the Nernst approximation was used to formulate the ion fluxes J_i across the boundary layers of the sample solution and of the membrane, respectively:

$$J_i = \frac{D_i'}{\delta'} (a_i - a_i')$$ (9)

$$J_i = \frac{D}{\delta} \frac{X}{|z|} (x_i' - x_i^o)$$ (10)

Analogous expressions hold for the ion j. X is the site density in the boundary layer (thickness δ) of the membrane, x_i' is the fraction of sites occupied at the membrane surface by the primary ion,

$$x_i' = \frac{a_i'}{a_i' + K_{ij} a_j'} ,$$ (11)

and x_i^o is the corresponding value for the bulk of the membrane. Using the zero-current condition, $J_i + J_j = 0$, the following fundamental result is obtained, given here in a generalized form:

$$a_i' + K_{ij} a_j' = a_i + K_{ij}^{Pot}(app.) a_j$$

$$= \frac{1}{2} (a_i + K_{ij}a_j - K_{ij}x_i^o C - \frac{D_j'}{D_i'} x_j^o C)$$

$$+ \frac{1}{2} \sqrt{(a_i + K_{ij}a_j - K_{ij}x_i^o C - \frac{D_j'}{D_i'} x_j^o C)^2 + 4 K_{ij}C (a_i + \frac{D_j'}{D_i'} a_j)}$$ (12)

$$\text{with } C = \frac{D}{D_j'} \frac{\delta'}{\delta} \frac{X}{|z|} .$$ (13)

Jyo and Ishibashi [8, 9] considered the special case where a membrane prepared or preconditioned with the primary ion ($x_i^o = 1$) is exposed to pure solutions of an interfering ion

$(a_i = 0)$. According to Eq. (12), ideal selectivity behavior is found only for sufficiently high sample activities, while an apparent selectivity coefficient of approximately unity is predicted for very diluted solutions:

$$E = E_i^o + s \cdot \log [K_{ij} a_j], \text{ for } a_j \gg C \text{ and } K_{ij} a_j \gg \frac{D_j'}{D_i'} C \qquad (14a)$$

$$E = E_i^o + s \cdot \log [\frac{D_j'}{D_i'} a_j], \text{ for } a_j \ll C \text{ and } \frac{D_j'}{D_i'} a_j \ll K_{ij} C \qquad (14b)$$

This means that a region of transient selectivities must appear at intermediate activities where the e.m.f. response curve for any ion other than the primary ion will exhibit a nonlinear section. A discontinuity of the response function is expected to occur at $a_j = C$ for ions preferred by the electrode ($K_{ij} \gg 1$), and an extended region with half of the Nernstian slope for ions rejected by the sensor ($K_{ij} \ll 1$). These theoretical predictions qualitatively agree with experimental findings, as is nicely demonstrated in Fig.1. Evidently, the incorporation into the membrane of the ion to be measured is a prerequisite for a linear response.

The variations of the apparent selectivities documented in Fig.1 may elucidate the reasons for the low precision or poor reproducibility of selectivity specifications found in the literature. One may even come to the unorthodox conclusion that many of the modifications or improvements claimed in the literature for liquid ion-exchange membrane electrodes did not really involve systematical changes in ion selectivity, but were largely a consequence of diffusion-induced artifacts. Figure 2 gives some support to this opinion. Apparently, the selectivity data reported for different chloride electrodes and different activities can be rationalized perfectly by assuming a single set of basic selectivity coefficients and assuming individual values for the experimental parameter a_x/C.

A more rigorous model of liquid-membrane electrodes shows that the pivotal parameter C is actually time-dependent. The quantity δ, as entering in Eqs. (10) and (13), roughly corresponds to the mean free path of diffusion in the membrane and has to be replaced by $\delta = \sqrt{\pi D t}$ [1]. From this, a function of the type $C = 1/\sqrt{t/\tau^*}$ is obtained. Hence the initial response behavior immediately after a sample activity change is characterized by a value of $C \to \infty$. Equation (12) then leads to the relation:

$$E = E_i^o + s \cdot \log \frac{K_{ij}(a_i + \frac{D_j'}{D_i'} a_j)}{K_{ij} x_i^o + \frac{D_j'}{D_i'} x_j^o} \qquad (15)$$

271

A similar result was presented by Hulanicki and Lewenstam [13] for solid-state membrane electrodes. These authors assumed that the pure salts of $\underline{i}$ and $\underline{j}$ coexist on the membrane surface, which implies that $x_i^o = 1$ and $x_j^o = 1$. The present formulation is also applicable to mixed solid phases or mixed adsorption isotherms for which $x_i^o + x_j^o = 1$. Equation (15) indicates that an initial selectivity factor of $D_j'/D_i' \sim 1$ should be observable. Thus, liquid- or solid-state membrane electrodes tend to respond to changes in total ion activity rather than to selectivity-weighted activity changes as long as the electrode and the bathing solution are far from reaching an equilibrium.

The long-time response behavior of ion sensors, on the other hand, is dominated by the bulk membrane selectivity. Indeed, for $C \to 0$, expansion of Eq. (12) into a series yields:

$$a_i' + K_{ij}a_j' = a_i + K_{ij}a_j + (K_{ij} - \frac{D_j'}{D_i'})(\frac{a_i}{a_i + K_{ij}a_j} - \frac{a_i^o}{a_i^o + K_{ij}a_j^o}) \cdot C$$

(16)

$$\cong a_i + K_{ij}a_j$$

Here the term x_i^o has been substituted in analogy to Eq. (11). Recalling the square-root time dependence of C, we then can use the first relation in Eq. (16) to assess the dynamic response of liquid-membrane electrodes in the presence of primary and interfering ions. The result conforms to an expression derived previously [1]:

$$E(t) = E(0) + s \cdot \log [A - B \cdot \frac{1}{\sqrt{t/\tau^*}} - (A-1) \cdot e^{-t/\tau'}]$$

(17)

where $E(0) = E_i^o + s \cdot \log [a_i^o + K_{ij}a_j^o]$; A and B are coefficients defined by Eqs. (8a), (16), and (17). An exponential term was added to account for diffusional relaxation in the aqueous boundary layer. In Fig. 3 the experimental e.m.f. vs. time profiles of a Ca^{2+} liquid ion-exchanger electrode in the presence of Mg^{2+} ions are fitted by computed values. The agreement between theory and experiment is remarkable and corroborates the basic diffusion model discussed in the present work.

TRANSIENT RESPONSE PHENOMENA

One interesting feature of Eq. (12) is its capacity to simulate transient responses or potential overshoots. According to Figs. 3 and 4, response delay or sluggish response is observed when primary ions are added to samples having a constant background of interferents with $K_{ij} \ll 1$ [6, 14]. Conversely, addition of the same interfering ions to solutions of the primary ion should lead to e.m.f. transients [14, 15] (see Fig. 4). Potential overshoots result from superimposition of the two effects. The reason for all the transient response phenomena illustrated in Fig. 4 is the diffusion-induced, time-dependent variation of the apparent potentiometric selectivity coefficient. This conforms to the tenor of an earlier theory [16] in which the distinction was made between surface selectivities, as reflected by the initial e.m.f.-excursion, and bulk membrane selectivities dominating the long-time response of ion sensors.

REFERENCES

1. W. E. Morf, The Principles of Ion-Selective Electrodes and of Membrane Transport, Akadémiai Kiadó, Budapest, 1980.
2. W. E. Morf and W. Simon, in Ion-Selective Electrodes in Analytical Chemistry (H. Freiser, ed.), Plenum, New York, 1978, p. 211.
3. W. E. Morf, E. Lindner, and W. Simon, Anal. Chem. 47, 1596 (1975).
4. E. Lindner, K. Tóth, E. Pungor, W. E. Morf, and W. Simon, Anal. Chem. 50, 1627 (1978).
5. P. L. Markovic and J. O. Osburn, AIChE J. 19, 504 (1973).
6. B. Fleet, T. H. Ryan, and M. J. D. Brand, Anal. Chem. 46, 12 (1974).
7. A. Hulanicki and R. Lewandowski, Chemia Analityczna 19, 53 (1974).
8. A. Jyo and N. Ishibashi, private communication, 1978 (publication in preparation).
9. for an appreciation of the Jyo-Ishibashi theory, see [1], pp. 245-57.
10. J. W. Ross, in Ion-Selective Electrodes (R. A. Durst, ed.), National Bureau of Standards Special Publication 314, Washington, 1969.
11. M. Oehme, Dissertation ETH, Juris, Zürich, 1977; K. Hartman, S. Luterotti, H. F. Osswald, M. Oehme, P. C. Meier, D.Ammann, and W. Simon, Mikrochim. Acta 1978II, 235.
12. P. C. Meier, D. Ammann, W. E. Morf, and W. Simon, in Medical and Biological Applications of Electrochemical Devices (J. Korýta, ed.), Wiley, Chichester, 1980.
13. A. Hulanicki and A. Lewenstam, Talanta 23, 661 (1976); 24, 171 (1977).
14. G. A. Rechnitz and G. C. Kugler, Anal. Chem. 39, 1682 (1967).
15. G. A. Rechnitz, in Ion-Selective Electrodes (R. A. Durst, ed.), National Bureau of Standards Spec. Publ. 314, Washington, 1969.
16. W. E. Morf, Anal. Lett. 10(2), 87 (1977).

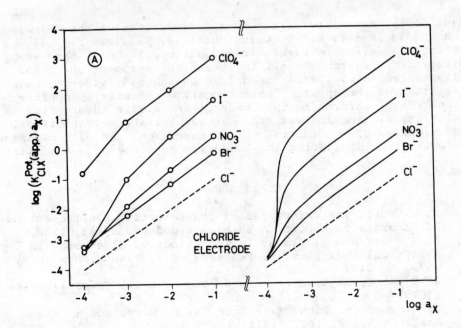

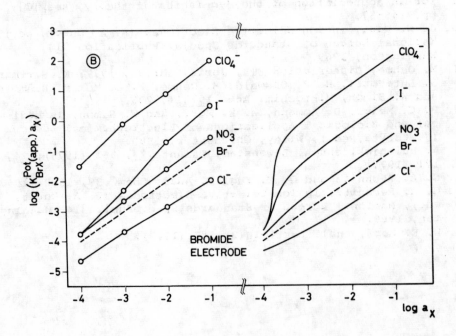

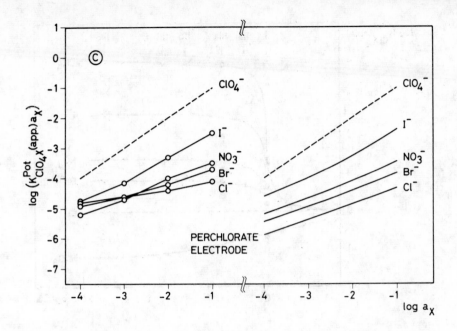

Figure 1. Response pattern of anion-sensitive liquid membrane electrodes based on quaternary ammonium salts (Aliquat 336S). A: chloride electrode, B: bromide electrode, C: perchlorate electrode. The shown functions are related to the apparent calibration plots for primary ions (dashed lines) and interfering ions (solid curves). Left traces: illustration of experimental selectivities, as given in table 1 of Ref. 7. Right traces: corresponding theoretical curves, calculated from Eq. (12) using $C = 1.8 \cdot 10^{-4} M$, $D'_j/D'_i = 1$, and $K_{ij} = k_j/k_i$ with $k_{ClO_4}:k_I:k_{NO_3}:k_{Br}:k_{Cl} = 10000: 400: 25: 7: 1$ [1].

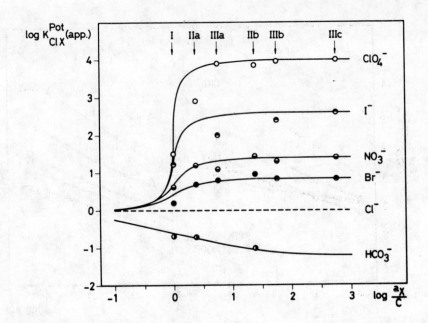

Figure 2. Reported selectivities for different chloride electrodes in comparison with theoretical expectations (Figure 1).
I: liquid membrane electrode (filter paper matrix) based on di-
methyldistearyl ammonium ions [10]. II: PVC membrane electrode
based on methyltridodecyl ammonium chloride; (a) 10^{-2}M, (b)
10^{-1}M solutions [11, 12]. III: PVC membrane electrode based on
methyltricapryl ammonium chloride (Aliquat 336S); (a) 10^{-3}M,
(b) 10^{-2}M, (c) 10^{-1}M solutions [7].

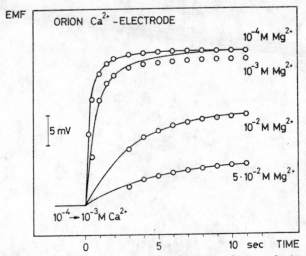

Figure 3. EMF-response vs. time profiles of a calcium-selective liquid membrane electrode (Orion 92-20) after an activity step $10^{-4}M \rightarrow 10^{-3}M$ Ca^{2+} in the presence of various activities of Mg^{2+}. The experimental curves are taken from fig. 2 in Ref. 6 (s = 25.5 mV). The points were computed from Eq. (17) with D_{Mg}/D_{Ca} = 1, K_{CaMg} = 0.011, τ' = 0.8s, and $\tau^* = 6.25 \cdot 10^{-6}$ M^2s[1].

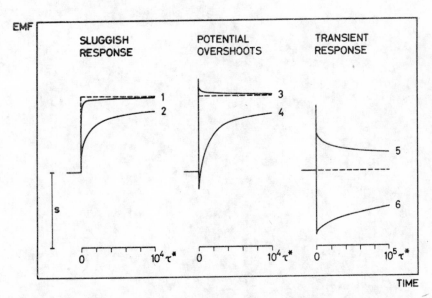

Figure 4. Different types of transient response phenomena. The curves were calculated from Eq. (12) with $C = (t/\tau^*)^{-1/2}$, D_j'/D_i' = 1, and K_{ij} = 0.01. The activity changes for ion i were $5 \cdot 10^{-4} \rightarrow 5 \cdot 10^{-3}M$ (curves 1-4) and $5 \cdot 10^{-4} \rightarrow 5 \cdot 10^{-4}M$ (5 and 6), respectively. The changes for ion j were $10^{-3} \rightarrow 10^{-3}M$ (1), $10^{-2} \rightarrow 10^{-2}M$ (2), $10^{-3} \rightarrow 2 \cdot 10^{-2}M$ (3), $10^{-2} \rightarrow 10^{-3}M$ (4 and 6), and $10^{-3} \rightarrow 10^{-2}M$ (5).

CHLORIDE AND CYANIDE DETERMINATION BY USE
OF THE FLOW-INJECTION METHOD USING ION-SELECTIVE FLOW-TYPE
ELECTRODES

H. MÜLLER

Sektion Chemie, Karl-Marx-Universität, Analytisches Zentrum, Leipzig, GDR

ABSTRACT

A method has been developed for the determination of chloride
and cyanide by use of the flow-injection method using ion-
selective flow-type electrodes.

INTRODUCTION

Among the current developments of mechanized microanalytical
techniques the flow-injection method holds a dominating
position /1-4/. The principle of the flow-injection analysis
consists in the injection of a sample into a constantly
flowing non-segmented carrier solution. For detection spectrophotometric techniques are most frequently applied. In recent
years, however, electro-analytical detection principles have
been applied more and more widely /5,6/.
The use of ion-selective electrodes frequently involves the
application of a flow-through cell. Examples of the construction of such flow-type electrodes are given e.g. by
Pungor et al. /5/, Ruzicka et al. /6/, Slanina et al. /7/.
All these types work in connection with a reference electrode (silver/silver chloride; calomel). Another principle
was chosen by us. The detector developed by us consists of
a combination of two ion-selective solid-state electrodes
of identical membrane material, both with direct internal
contact. The arrangement enables measurement of a difference
in potential between the surfaces of the membranes.

The detector was installed in a system which is shown in figure 1. Chloride concentrations up to 10^{-5}M and cyanide concentrations up to 10^{-6}M were determined in micro-samples.

EXPERIMENTAL

Figure 2 shows the construction of our flow-through electrode. Ionic connection is provided subsequent to the electrode. The received signal is amplified and recorded.

E.m.f. measurements were made with a precision pH-meter type MV 87 (VEB Präcitronic, Dresden, DDR) in connection with a recorder type "endim" (VEB Meßapparatewerk Schlotheim, DDR). The samples were injected by means of a sample loop (volume 70 µl) or with a syringe via a septum sample injector. The mixing of the samples with the carrier solution was carried out in a reaction coil (50 cm long).

RESULTS

If the concentrations of both solutions (see fig. 2) are identical, a potential difference close to zero is measured. If a sample is injected, a difference in concentration and thus a potential difference between the two membrane surfaces arises, which is subsequently measured.

Chloride determination

All studies were carried out with a flow-through cell on silver sulphide/silver chloride basis. The response of the electrodes to chloride is shown in figure 3.

Up to 10^{-4} M chloride the response is near Nernstian; the difference between 10^{-4}M and 10^{-5}M, however, amounts to not more than 30 mV. At constant flow rate (21 ml min^{-1}) the influence of sample volume is shown in figure 4. It follows that any point of the rising part of the peak is as good a measure of the analyte concentration as the "steady state" value eventually reached and it is therefore a waste of sample and carrier solution and of time to wait for plateau forming in the response curve.

Figure 5 shows the dependence of the peak height on the concentration of the injected chloride solution.

The reproducibility of the technique is shown in table 1. Various foreign ions were tested in order to study their influence on the determination of chloride. The results showed that under flow conditions the disturbances are larger than under non-dynamic conditions.

Chloride determination in blood serum

The natural range of chloride in serum is near 0.1 molar (9.6 to 10.4 . 10^{-2}M Cl$^-$). The test conditions were optimised in order to facilitate distinction between these two extreme values. Figure 6 shows that a difference in potential of about 5 mV occurs between the two limiting values. When the samples were directly applied, then the relative standard deviation is about 2.5%. For calculation of the error caused by the matrix (e.g. influence of SH-groups of the albumen) control studies were made on dealbumenized samples. De-albumenization was effected by means of 2 M nitric acid. After centrifuging the electrolyte solutions were neutralized with sodium hydroxide and buffer solution with pH 4.6 before measuring.

The deviations of our results related to the standard serum Precinorm S (10.00 . 10^{-2}M Cl$^-$) were -6% for de-albumenized and +11% for non-treated serum. The minus values may be due to losses during de-albumenizing and the higher values due to matrix influence, but this problem will be further studied. By application of millipore membrane filters and by sample conditioning it is tried to remove the matrix influence.

Studies on cyanide determination

The studies were carried out with a solid-state flow-through electrode (Ag$_2$S-membrane) working on the difference principle described above. Using a 10^{-4} M silver nitrate solution with an ionic strength of 0.1 (KNO$_3$), which was adjusted to pH 11 with ammonia, cyanide solutions between 10^{-6} and 10^{-3} M were analyzed (sample volume 70 μl, sample loop). The silver ions not consumed in the reaction with the cyanide ions are measured. The standard deviation was about 5%.

DISCUSSION

With the flow-type electrode which was developed by us and installed in a flow-injection system it is possible to determine chloride up to 10^{-5}M and cyanide up to 10^{-6}M in micro-samples (10-70 µl) with a good reproducibility. First experiments have shown that it is possible to determine chloride in human blood serum with a very simple technique.

REFERENCES

1. G.Nagy, Zs.Fehér and E.Pungor, Anal.Chim.Acta 52,47(1970)
2. J.Ruzicka and E.H.Hansen, Anal.Chim.Acta 78,145(1975)
3. Betteridge,D., Anal.Chem. 50,832A(1978)
4. J.Ruzicka and E.H.Hansen, Anal.Chim.Acta 99,37(1978)
5. E.Pungor, Zs.Fehér, G.Nagy, K.Tóth, G.Horvai and M.Gratzl Anal.Chim.Acta 109,1(1979)
6. J.Ruzicka, E.H.Hansen and E.A.Zagatto, Anal.Chim.Acta 88,1(1977)
7. J.Slanina, W.A.Lingerak and F.Bakker, Anal.Chim.Acta 117,91(1980)

Table 1. Reproducibility of the determination of chloride (v_r in %; $c_0=10^{-5}$M Cl$^-$; ionic strength 0.1 M KNO$_3$; N=10)

Injection by means a sample loop (volume 70 µl) :

	concentration of the injected samples, M		
	10^{-4}	10^{-3}	10^{-2}
1.day	8.3	4.2	4.2
2.day	9.8	3.1	5.5
from day to day	10.0	8.5	8.8

Injection with a syringe via a septum sample injector :

1.day	1.4	1.2	0.4
2.day	4.0	1.4	0.6
from day to day	11.0	4.1	3.6

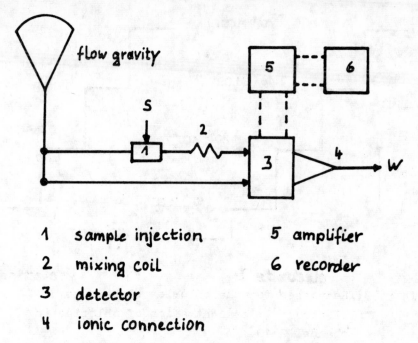

flow gravity

S

1   2   3   4   5   6   W

1   sample injection   5   amplifier

2   mixing coil   6   recorder

3   detector

4   ionic connection

Figure 1: Flow injection manifold for the difference
potentiometric determination of chloride

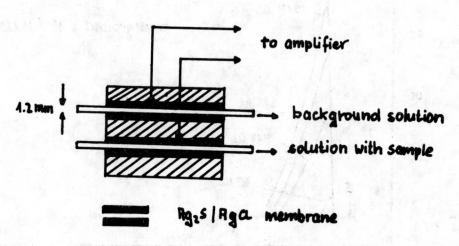

to amplifier

1.2 mm

background solution

solution with sample

$Ag_2S/AgCl$ membrane

Figure 2: Ion-selective flow-type electrode

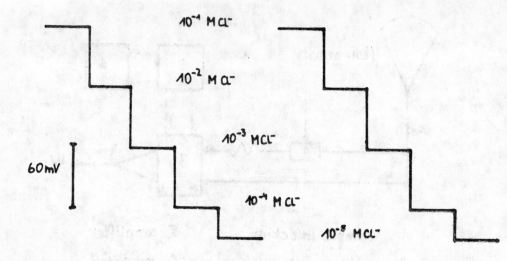

Figure 3: Responses for the chloride electrodes
electrode I: concentration of chloride in channel 2 10^{-5} M
electrode II: concentration of chloride in channel 1 10^{-5} M

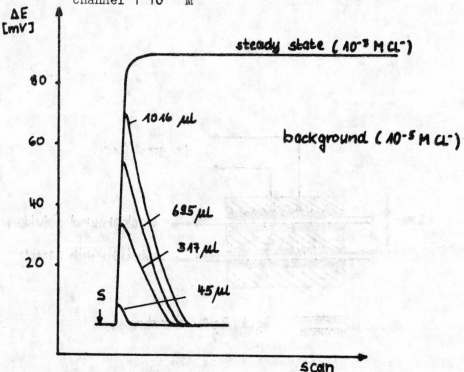

Figure 4: Dependence of signal from the volume of the injected sample (10^{-3}M Cl$^-$)

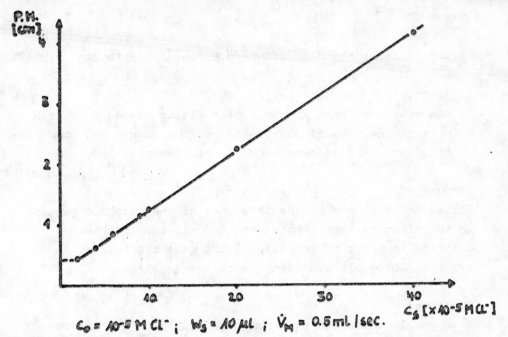

$C_0 = 10^{-5} M\ Cl^-$; $W_S = 10\ \mu L$; $\dot{V}_M = 0.5\ mL\ /sec.$

Figure 5: Dependence of the peak height (P.H.) from the
concentration of the injected samples (2 ... 40
$10^{-5}M\ Cl^-$)

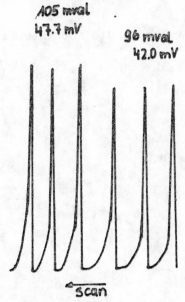

105 mval
47.7 mV

96 mval
42.0 mV

← scan

Figure 6: Peaks of chloride ion standards (9.6 and 10.5
. $10^{-2}M\ Cl^-$)

QUESTIONS

Participants of the discussion: W.Simon, J.D.R.Thomas, H.Müller

Question:
It was mentioned that an error of + 11% is caused by matrix
effects in the case of non-treated serums. Could you comment
on this fact ? Is it some kind of interference ?

Answer:
The error is caused by the influence of the sulphide groups
in the albumin. Experiments were made in the absence of
chloride in order to investigate the effect of serum. An
error of 10 to 11% was found in this case also.

Question:
Some people commented that they were unable to use the flow
injection analysis for monitoring urine, because of the
variation in the viscosity of urine. Could you comment on
this fact as you mentioned urine as an example ?

Answer:
The viscosity of the sample did not interfere in the course
of our experiments.

RESPONSE TIME STUDIES ON SOLID-STATE ION-SELECTIVE ELECTRODES WITH INTEGRAL ELECTRONICS

K. NAGY AND T. A. FJELDLY*

Division of Applied Chemistry SINTEF, The University of Trondheim,
The Norwegian Institute of Technology, N-7034 Trondheim, Norway
*Electronics Research Laboratory, The University of Trondheim, The Norwegian Institute
of Technology, N-7034 Trondheim, Norway

ABSTRACT

In this report we discuss response time measurements on ion-selective electrodes with integral electronics. We have investigated electrodes with various silver salt membranes and, in particular, a new LaF_3 electrode with a reversible solid-state inner contact. The initial response was measured with a rapid flow injection technique. For comparison, similar tests were also performed on commercial electrodes with internal reference solution. We find no significant difference in the behaviour of the present and commercial fluoride electrodes. In both cases there is a tendency towards longer response times with decreasing fluoride concentration in the injected solution. The same trend is found in total response time measurements performed in a practical analytical set-up.

INTRODUCTION

The applicability of ion-selective electrodes in chemical analysis depends on such electrode properties as sensitivity, selectivity, stability, response time and life time. The study of response time is important from the point of view of understanding the electrode processes, but it is also of great practical interest in the terms of efficiency and economy in the laboratory.

In the past, the literature on electrode response time has largely been restricted to experimental work, and discussions on phenomenological grounds. However, notable exceptions are the work by Buck on the transient behaviour of glass electrodes [1,2], and his elucidation of the general subject matter in recent review articles [3,4].

Experimentally, response time can be measured in different configurations. In the fast-flow technique one attempts to create a step activity change at the external membrane surface. Rapid flow past the membrane tends to reduce the thickness of the stagnant surface film, which causes a reduction in the transport time associated with the film diffusion.

A variation on this method is the injection technique, where fresh solution is rapidly injected into a narrow gap between the reference and the ion-selective electrode, displacing a small drop of initial solution. In the dipping experiment the ion-selective electrode is suddenly immersed in the test solution; but in this case there is no initial reference activity with which to compare.

From the point of view of practical analysis, the slower relaxation processes are of primary interest. They govern the times needed to reach analytically significant stability in practical samples. Such properties are conveniently studied, for example, by injecting affecting ions into rapidly stirred buffered solutions. Such measurements will, of course, also involve relaxation processes associated with homogenization and chemical equilibration in the solution bulk.

In the present work we report measurements on the rapid electrode response by the injection technique for electrodes based on solid-state pressed pellet membranes (silver salts) and, in particular, LaF_3 single crystals. For the latter, we have also studied the overall response time in a practical analytical set-up. The electrodes used are of a new all-solid-state construction with integral microelectronic preamplification circuitry [5-7]. A number of such electrodes with pressed pellet membranes of Ag_2S, $AgI-Ag_2S$, $AgCl-Ag_2S$, $PbS-Ag_2S$ and single crystalline LaF_3 have been manufactured and tested [5-9]. In Fig. 1 is shown the basic construction of the electrodes. The single chip operational amplifier operates in a voltage follower mode to give a pure impedance transformation at close proximity to the membrane, in order to minimize noise pick-up. A reversible, solid-state inner contact has been realized by using a $Ag-AgF$ layered structure at the inner membrane surface. Details on the fabrication and properties of this reversible, all-solid-state fluoride electrode are published elsewhere [9].

The main objective of this work has been to supplement previously reported properties of these new electrodes in order to establish the applicability of the basic electrode construction. Thus, the emphasis has been placed on comparison with established, commercial electrodes of conventional construction. For this purpose the fluoride electrode was chosen, partly because of the newly developed solid-state contact, and partly due to the scarcity of literature on response time measurements on the LaF_3 electrode.

EXPERIMENTAL

The initial response time of the electrodes was investigated with an experimental set-up as shown in Fig. 2. The ion-selective electrode and a reference electrode were pointed against each other with a separation of about 1 mm, and samples of test solution were injected into this gap with a syringe. The electrode response was recorded by means of a Tektronix storage oscilloscope, type 549. The oscilloscope sweep was triggered by forcing the test solution past a separate electrode pair positioned just in front of the electrodes of interest, thereby producing a voltage step by shorting the trigger electrodes. The delay between the trigger and the signal could be adjusted by varying the distance between the electrode pairs.

For the total response time measurements we used 50 ml volumes of buffer/standard solution. The solutions were stirred by means of a magnetic stirrer. Microburettes and pipettes were used for providing known additions of standards to the test solutions.

The silver salt membranes were pressed from homogenized, pulverized mixtures of silver salts. The dimensions of the membranes used were about $2.5 \times 2.5 \times 0.5$ mm^3. For the fluoride electrode we used single crystalline LaF$_3$ membranes of cylindrical shape, 5 mm in diameter and 2 mm thick, provided by BDH Chemicals Ltd. The reference electrode was a Radiometer double junction calomel electrode, model K701, unless otherwise noted.

RESULTS AND DISCUSSION

In the measurements of response time by the injection technique, it was convenient to use relatively large activity changes. Such measurements do not easily lend themselves to interpretation in terms of linearized theory [3,4], but the experimental conditions are easily reproduced for the purpose of comparing different electrodes.

Fig. 3 shows typical responses for two silver salt electrodes, Ag$_2$S and AgI-Ag$_2$S, as obtained by the injection technique. In the case of the Ag$_2$S electrode a small drop of diluted AgNO$_3$ ($\sim 10^{-5}$ M) served as an initial solution. This was displaced by injection of 0.1M AgNO$_3$ into the gap (see Fig. 2). An activity change from 10^{-4}M to 0.1M KI was used with the AgI-Ag$_2$S electrode. A momentary electrical contact between the trigger and the measuring electrodes probably was the source of the transient observed at the onset of the time response.

By assuming a simple exponential relaxation towards new equilibrium we find that the response time for these electrodes usually ranges between 10 to 20 ms. This is somewhat better than results obtained by others on conventional electrodes with internal solution contact [10-12]. It is believed that the silver salt electrodes have a rapid surface ion exchange and that no diffusive, hydrolyzed surface layer exists. Likewise, the space-charge relaxation time associated with the membrane resistance and the double-layer capacitance, has been found to be an order of magnitude smaller than the present values [4]. Thus, the observed relaxation is probably associated with diffusive transport in the stagnant solution film outside the membrane. Morf et al [13] have shown that the response vs. time profile for this case varies according to

$$\phi(t) - \phi(\infty) = \frac{RT}{F} \ln \left\{ 1 - (1 - \frac{a(o)}{a(\infty)}) e^{-\tau/t} \right\} \tag{1}$$

where

$$\tau \sim \delta^2 / 2D \tag{2}$$

a(o) is the initial and a(∞) the final activity of affecting ions, δ is the film thickness, and D is the diffusion coefficient of the ions. It is easily shown that eq. (1) reduces to simple exponential relaxation at long times.

A more extensive investigation of the initial response was carried out on fluoride electrodes, both on the all-solid-state microelectronic version and on a commercial electrode with internal filling solution (Orion 94-09). The time response was measured for different activity step changes using a

TISAB buffer solution. The activity changes were always from 0.005g F/1 ($\sim 2.5 \times 10^{-4}$ M) to 0.05, 0.5 or 5 g F/1. The results are shown in Fig. 4 for the all-solid-state electrode and for the commercial electrode. Similar results were also obtained with a 2.5M HCl buffer solution.

One immediate commentary, with bearing on the objective of this investigation, is that there seems to be no significant difference in the behaviour of the two electrodes. Both give rise to response times in the range 20-100 ms, depending on the experimental circumstances. Most likely, therefore, the observed responses are associated with relaxation processes at the outer membrane surface, and not with details of the electrode construction. Another point of interest is that the present relaxations are slower and seem to be more complex than for the silver salt electrodes. The experiments indicate that for large activity changes there is initially a rapid relaxation which, at longer times, gives way to slower processes. For smaller activity changes, i.e. less than two decades, the slower decay seems to dominate all the way. Also, instead of approaching steady state, there seems to be a residual potential drift with a time constant that is long on the present time scale. It appears, therefore, that there are mechanisms at work in addition to the solution layer diffusion discussed in connection with the silver salt electrodes. One possibility is that the LaF_3 possesses a hydrolyzed surface film, as suggested by Veselý [14], which may tend to slow down the ion exchange. Our experiments on the admittance behaviour of LaF_3 with solution contacts [9] are consistent with the assumption of a gel layer at the LaF_3-solution interface.

TOTAL RESPONSE TIME IN PRACTICAL ANALYSIS

Our all-solid-state fluoride electrode and a commercial electrode (Orion 94-09) were also compared in a practical, analytical set-up as described earlier. NaF standards were injected in succession into TISAB buffer solutions, and the potential response was recorded on a x-t recorder (Watanabe Servo Recorder, type 652). Typical results are shown in Fig. 5. As a general trend, we find that the total response time for one-decade fluoride activity steps decreases as the concentration increases. It also appears that the $Ag-AgF-LaF_3$ electrode and the Orion 94-09. lie approximately within the same range of speed.

In a more stringent test, the response time was also measured with a computerized analyzer system, the SINTALYZER [15]. This system is designed for potentiometric analysis with multiple addition and titrimetric techniques. It has a stability criterion where the electrode potential-reading is accepted only when the potential drift is less than 0.1mV over a specified lapse of time, typically 10 or 20 seconds.

Table 1 shows the acceptance times found for one-decade increases in fluoride concentration, both with the present electrode and with a commercial combination electrode (Orion 96-09). The reference electrode of the latter was here used for both cases. Note that the times measured include both homogenization time and the residual time set by the acceptance criterion. Again, this test shows that the present solid-state electrode and the commercial electrode are within the same range of speed, with the present electrode perhaps a little faster at low concentrations.

In Table 2 are finally shown analytical results and total analysis time for six-point multiple addition routines, using the SINTALYZER and the Ag-AgF-LaF$_3$ electrode. The acceptance criterion was 0.1mV potential stability for 10 seconds. Note that the total analysis time now includes the injection time for five additions of standard, which typically counts for 20-30% of the total time.

SUMMARY

The response time for new all-solid-state ion-selective electrodes with integral electronics was studied both by fast flow injection technique and in a practical analytical situation. By the fast injection method we found response times in the range 10 - 20 ms for electrodes with silver salt membranes, probably limited by diffusive transport in the stagnant solution film outside the membrane. In similar experiments on commercial and all-solid-state fluoride electrodes we found response times between 20 - 100 ms, depending on the activity change. The relaxation behaviour was mor complex in this case, which possibly may be attributed to the presence of a hydrolyzed surface film (gel) on the LaF$_3$ surface. Using the fluoride electrodes in a practical, analytical set-up we found total response times in the range 15 - 50 seconds. The response time increased somewhat with decreasing fluoride concentration. There was no significant difference in the performance of the present and commercial electrodes.

In conclusion, the experiments indicate that the electrodes with reversible solid membrane contacts and conventional electrodes with internal filling solution have a comparable performance [7]. However, the all-solid-state version offers advantages in terms of miniaturization, ruggedness and applicability (i.e. temperature and pressure range). Also, the solid contacts are more compatible with microelectronic manufacturing techniques.

REFERENCES

[1]   R.P. Buck, J. Electroanal. Chem., 18, 363 (1968), and ibid., 18, 381 (1968).

[2]   R.P. Buck and I. Krull, ibid., 18, 387 (1968).

[3]   R.P. Buck, in Ion-Selective Electrodes, E. Pungor, Editor, Akadémia Kiado, Budapest (1978).

[4]   R.P. Buck, in Ion-Selective Electrodes in Analytical Chemistry, Vol. 1, H. Freiser, Editor, Plenum Press, New York (1978).

[5]   T.A. Fjeldly and K. Nagy, SINTEF report STF21, 79009 (1979).

[6]   T.A. Fjeldly, K. Nagy and J.S. Johannessen, J. Electrochem. Soc., 126, 793 (1979).

[7]   T.A. Fjeldly and K. Nagy, this conference.

[8]   K. Nagy, E. Kleven, T.A. Fjeldly and D. Fremstad, Z. Anal. Chem., 295, 362 (1979).

[9]   T.A. Fjeldly and K. Nagy, J. Electrochem. Soc., 127, 1299 (1980).

[10] K. Toth and E. Pungor, Analytica Chimica Acta, 64, 417 (1973).

[11] R. Rangarajan and G.A. Rechnitz, Analytical Chemistry, 47, 324 (1975).

[12] T.H. Ryan and B. Fleet, Proc. Analyt. Div. Chem. Soc., 53 (Feb. 1975).

[13] W.E. Morf, E. Lindner and W. Simon, Analytical Chemistry, 47, 1598 (1975).

[14] J. Veselý, Electroanal. Chem. Interfacial Electrochem. 41, 134 (1973).

[15] K. Nagy and E. Keul, Proc. Meeting of the Metallurg. Soc. of AME, Denver 1978, paper no. A78-39.

Table 1. Response times of Ag-AgF-LaF$_3$ and Orion 96-09 fluoride electrodes measured with the SINTALYZER system, for successive one-decade increases in fluoride concentration. Acceptance criterion: 10 and 20 seconds of potential stability to within 0.1mV. Buffer solution: 1 M HCl. Sample volume: 50 ml. Fluoride solutions: Pure NaF standards.

Electrode	Fluoride conc. jump (mole/ℓ)	Acceptance criterion (sec.)	
		10 sec.	20 sec.
Ag-AgF-LaF$_3$ (AE 8003)	$0 - 10^{-6}$	32	43
	$10^{-6} - 10^{-5}$	32	43
	$10^{-5} - 10^{-4}$	19	34
	$10^{-4} - 10^{-3}$	16	25
	$10^{-3} - 10^{-2}$	17	29
ORION 96 - 09	$0 - 10^{-6}$	53	119
	$10^{-6} - 10^{-5}$	26	62
	$10^{-5} - 10^{-4}$	17	32
	$10^{-4} - 10^{-3}$	15	27
	$10^{-3} - 10^{-2}$	16	30

Table 2. Practical six-point standard addition analyses with the SINTALYZ-ER and the Ag-AgF-LaF$_3$ electrode. The analytical results are compared with the true concentrations of fluoride. Also shown are the acceptance times for the first potential reading and the total analysis time. Buffer solution: 1M HCl. Sample volume: 50 ml. Fluoride solutions: Pure NaF standards.

Fluoride conc. (ppm)		Anal. Time (min., sec.)	
True	Result of Analysis	1. Potential	Total
0.010	0.010	0^{33}	4^{54}
0.020	0.026	0^{47}	5^{05}
0.100	0.104	0^{53}	4^{22}
0.200	0.204	1^{03}	4^{21}
1.00	1.01	0^{54}	3^{14}
2.00	1.99	0^{45}	2^{54}
10.00	10.3	0^{47}	2^{30}
20.00	20.2	0^{45}	2^{28}

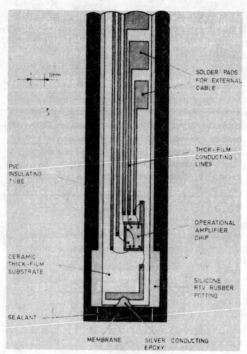

Fig. 1   Basic design of all-solid-state ion-selective electrode with
         integral amplifier electronics.

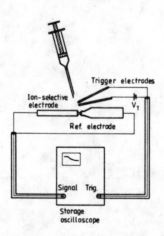

Fig. 2   Experimental set-up for initial response time measurements.

294

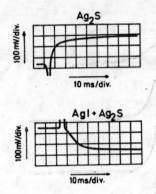

Fig. 3 Typical responses for two silver salt electrodes obtained with the injection technique.

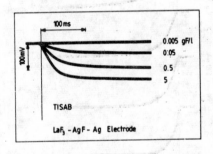

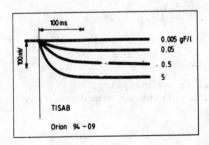

Fig. 4 Initial responses for the all-solid-state electrode and for Orion 94-09 obtained with the injection technique.

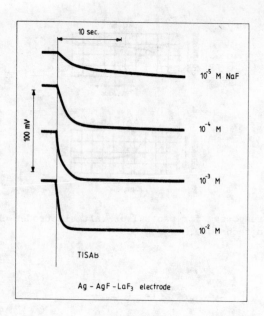

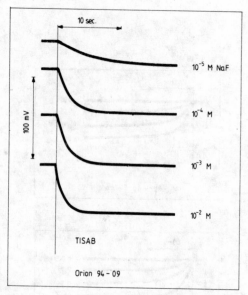

Fig. 5   Total response in an analytical set-up for one-decade increases
in fluoride concentration.   Sample volumes: 50 ml.

EQUILIBRIUM REACTIONS IN POTENTIOMETRIC ANALYSIS

G. PETHŐ AND K. BURGER

Department of Inorganic and Analytical Chemistry, L. Eötvös University, Budapest, Hungary

ABSTRACT

The application of a new evaluation method /1/ for the use of equilibria having low equilibrium constants in potentio-metric analysis is presented. The model reactions studied are: the protonation of weak bases /to be measured otherwise only in non-aqueous solution/, the silver precipitation of chloride ions in 10^{-4} M solution and the ion pair formation of bulky quaterner ammonium cations with tetrathiocyanato cobaltate/II/ anions.

INTRODUCTION

Equilibrium reactions which do not proceed quantitatively at titration equivalence points, e.g. reactions involving weak bases or weak acids, or metal complexes or precipitates with low equilibrium constants, can be utilized analytically by plotting the concentration of the reagent consumed /e.g. the bound proton or bound metal concentration, etc./ as a function of the volume of titrant. Saturation curves are thus obtained and the limiting value is equivalent to the concen-tration of the base /or ligand/ to be determined /1/. The concentration of bound proton /or bound metal/ is calculated from the difference between the total reagent concentration added and the concentration of the free protons /or uncomplexed metal ions/ measured potentiometrically.

The analysis is performed analogously to normal potentiometric titrations except that the ionic strength of all solutions must be kept constant by adding an inert salt /e.g., potassium nitrate/ and the volume of the solution at the beginning of the titration /v_o/ must be known. Constant ionic strength is necessary to ensure that the activity coefficients remain constant; the electrode can thus be calibrated to measure the concentration of free proton or uncomplexed metal ion /both represented by $[X]$ in the following/. Their total concentration /C_i/ is calculated after addition of v_i cm^3 of the standard solution of concentration C_X at each point of the titration from the equation

$$C_i = \frac{C_X v_i}{v_o + v_i},$$

as well as the concentrations of free X from the Nernst Law:

$$[X_i] = 10^{(E_i - E_o) z/0.059}$$

The E_o value is determined from e.m.f. measurements in solutions of identical ionic strength containing known concentrations of X.

For quantitative evaluation, the differences $C_i - [X]$ corrected for dilution as $(C_i - [X])(v_o + v_i)/v_o$ are plotted as a function of the additions v_i of the standard solution. The horizontal section of this curve is projected to the ordinate to obtain the concentration of bound X after the completion of the reaction, which is equivalent with the original concentration to be measured.

The theoretical background of this evaluation method based on Bjerrum's complex formation function /2/ is presented in /1/. Its practical application is illustrated on several examples in the present paper. The models used are the protonation equilibria of weak bases /otherwise determined by non-aqueous titrations/, the silver precipitate formation of chloride ion in dilute solutions and ion pair formation equi-

libria of bulky quaterner ammonium cations with tetrathiocyanato cobaltate/II/ anion.

EXPERIMENTAL

An Orion model 701 A digital pH meter was used for all titrations. In the pH-metric determinations an Orion 91-01-00
glass electrode was used for measurement of the free proton
concentration, in the argentometric titrations the free sil-
ver/I/ concentration was measured by a Radiometer P 4011 silver electrode, in the study of the ion pair formation reactions a membrane electrode containing a benzalkonium-tetra-
thiocyanato cobaltate/II/ ion pair as electroactive substance
incorporated into a polyvinyl chloride matrix was used. This
electrode was prepared as described previously /3/ and showed
a reversible Nernstian function in the cobalt concentration
interval of $10^{-1} - 10^{-4}$ M with a slope of 29.3 mV. In each
titration a Radiometer K 100 calomel electrode in a Wilhelm
bridge /4/ served as reference. The standard solution was
measured from a Radiometer ABU 12 automatic burette. The initial volume /v_o/ of titrations was 10.00 or 20.00 cm^3. The
ionic strength of the solutions for the pH-metric measurements
was adjusted to 1.0 with sodium chloride, or those used in the
chloride determinations to 0.5 with potassium nitrate and of
those used in the ion pair formation reactions to 3.0 with
potassium thiocyanate. The calculations needed for the construction of the curves of C - $[x]$ vs cm^3 of standard solutions were done with a programmable pocket calculator /TI 59/.
Such typical curves together with the primary mV vs cm^3 of
titrant curves are shown in Figs. 1.-3.

RESULTS AND DISCUSSION

The results of the measurements are collected together
with experimental error of the measurement and its standard
deviation in the Table.

RESULTS AND EXPERIMENTAL ERRORS OF THE ANALYSES

Compound	Standard solution	Calculated mgcm^{-3}	Measured mgcm^{-3}	Experimental error, rel. %	Standard deviation of error, %
Sulfacetamide – – sodium	0.5 M HCl	25.45	25.34	−0.43	± 0.50
	0.1 M HCl	25.45.	25.33	−0.48	± 0.62
	0.1 M HCl	2.545	2.553	+0.31	± 1.48
	0.01 M HCl	2.545	2.551	+0.24	± 0.39
	0.01 M HCl	0.2599	0.2606	+2.42	± 1.05
Aminopyrine	0.5 M HCl	22.82	22.58	−1.05	± 0.67
	0.1 M HCl	2.282	2.214	−2.99	± 0.63
	0.01 M HCl	0.2339	0.2354	+3.24	± 0.56
Nicotinamide	0.5 M HCl	23.66	23.05	−2.58	± 0.48
	0.5 M HCl	11.83	11.59	−2.01	± 0.53
	0.5 M HCl	5.915	5.594	−5.42	± 0.29
Potassium chloride	0.05 M AgNO$_3$	3.810	3.822	+0.30	± 0.29
	0.01 M AgNO$_3$	0.381	0.386	+1.42	± 0.43
	0.01 M AgNO$_3$	0.0381	0.0380	−0.39	± 1.01
Chloropyramine	0.1 M CoCl$_2$	0.345	0.343	−0.70	± 0.42
Benzalkonium chloride	0.1 M CoCl$_2$	0.322	0.314	−2.48	± 2.69

Typical experimental curves and the corresponding derived curves are shown in the Figures. Comparison of the primary conventional titration curves with those constructed on the basis of the new approach gives immediate proof of the superiority of the latter. None of the primary titration curves showed an inflection which could be evaluated as the endpoint of the titration by conventional methods. The analytical results got from the limiting values of the C - [X] vs cm^3 titrant curves indicate that analytical determinations which cannot be performed by the usual potentiometric titration in aqueous solution /e.g. determination of weak bases, of chloride in 10^{-4} M concentration and of quaterner ammonium cations in ion pair formation reactions/ can be measured using the new method with good accuracy.

REFERENCES

1. K.Burger, G.Pethő and B.Noszál, Anal. Chim. Acta <u>118</u>, 93 /1980/
2. J.Bjerrum, Metal Ammine Formation in Aqueous Solutions, Haase, Copenhagen, 1941
3. K.Burger and G.Pethő, Anal. Chim. Acta <u>107</u>, 113 /1979/
4. W.Forsling, S.Hietanen and L.G.Sillén, Acta Chem. Scand. <u>6</u>, 905 /1952/

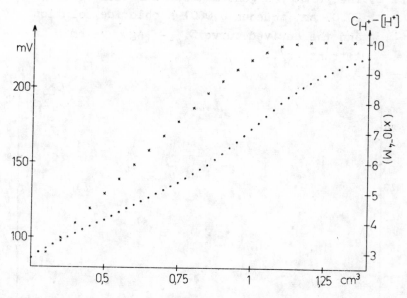

Figure 1. The primary potentiometric titration curve of 10.00 cm^3 of the 0.01 M aqueous solution of aminopyrine /·/ and the derived curve $C_H - [H^+]$ vs cm^3 of titrant /x/.

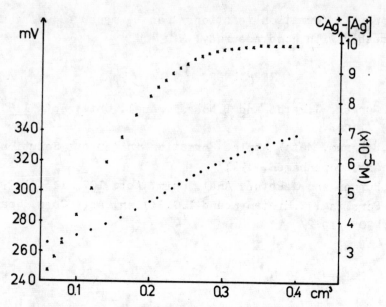

Figure 2. The primary argentometric titration curve of
20.00 cm³ aqueous 0.0001 M chloride solution /•/
and the derived curve C_{Ag} - $[Ag^+]$ vs cm³
titrant /x/.

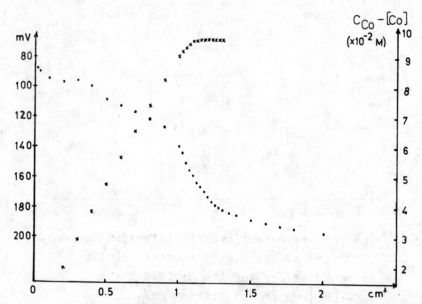

Figure 3. The primary titration curve of 10.00 cm³ aqueous
0.02 M benzalkonium chloride solution with 0.1 M
tetrathiocyanato cobaltate standard solution /•/
and the derived curve C_{Co} - $[Co]$ vs cm³ titrant
/x/.

302

QUESTIONS AND COMMENTS

Participants of the discussion: E.Pungor, A.Hulanicki,
K.Burger, G.Pethő

Question:
The method used, which you already published, is very interesting. The signals you got primarily could not be evaluated
directly, but after a transformation the results were quite
good. Was the error of the signals checked ? I think there
is an error propagation involved in the transformation.

Answer:
May be there is a misunderstanding. In the normal titration
there is a change in emf and the emf-values are used to get
the concentration of the free ligand, similarly to equilibrium
measurements in coordination chemistry. This way the free
ligand concentration is obtained from the emf-values for
every point of the titration, and the difference between the
total and free ligand concentration gives the free ligand
concentration. So we do not need the volume functions of the
titrant consumed, this is only used to correct for the
dilution during the measurement.
    We make the calculations from the values on the y axis from
the concentrations of the protons liberated and thus, the
accuracy is not determined primarly by the error of the emf
measurement.

Comment:
Maybe I misunderstood what you said. However, you have only
two data at hand: the amount of the reagent added and the
potential response.

Answer:
Yes; and the potential response is calibrated with respect
to the free ligand concentration. Therefore it is very
important to make all the titrations at the constant ionic
strength at which the calibration was carried out.

Question:

Are the results improved by the transformation you applied ?

Answer:

Yes, they are improved remarkably. If we use the Gran lineari-
zation, we can end up with an error of 10% or even greater
with such diluted solutions, whereas using the transformation
outlined, we have an error of 1.0-1.5%.

Question:

It was mentioned that very weak acids can be titrated by using
the method described. What is the limit of pK above which the
method can not be used ?

Answer:

The limiting pK value is 10-11. Acids or bases with this
pK value can still be titrated with an error of $\pm$10%.

Question:

It was mentioned that a cobalt-selective electrode was used
in the titrations. Was it proved that the electrode response
is linear in the concentration range involved ? What kind
of cobalt-selective electrode was used ?

Answer:

The electrode response was found to be linear in the range
10^{-2}-10^{-4} M. The active material of the electrode contained
tetracyano-cobalt/II/ ion. The scheme of the electrode was
shown in a slide.

ION-SELECTIVE ELECTRODES FOR DETERMINATION OF GOLD(I) AND SILVER IN THIOUREA SOLUTIONS

O. M. PETRUKHIN, YU. V. SHAVNJA, A. S. BOBROVA AND YU. M. CHIKIN

V. I. Vernadsky Institute of Geochemistry and Analytical Chemistry,
USSR Academy of Sciences, Moscow and Irkutsk State Scientific-Research Institute of Rare and Nonferrous Metals, USSR

INTRODUCTION

Thiourea complexes of gold/I/ and silver have been widely used to solve research and technological problems. Therefore, it is very important to elaborate a rapid and precise method for analysis of the solutions of this type.

Gold/I/ and silver form thiourea complexes in thiourea solutions. Different methods can be used for the determination of metals in solutions containing a complex-forming reagent. We used an approach analysis of the solutions based on synthesizing and using an electrode-active compound involving a metal complex which is present in the solution studied. It is advisable because the gold and silver thiourea /tu/ complexes have very high stability constants and in the presence of small excess of thiourea equilibria are coompletely shifted toward the complexes. We have investigated in detail the extraction of gold/I/ and silver from thiourea solutions to look for an ion-exchanger material and conditions for determination of gold and silver/1/.

EXPERIMENTAL

Gold/I/ and silver form cationic complexes in thiourea solutions and, therefore, we have investigated the extraction of these metals by nitrobenzene, 1,2-dichloroethane and chlorobenzene in the presence of high molecular weight anion-forming agents of different kinds: aliphatic carboxylic acids, Na-salt of naphthalene sulphonic acid, picric acid /Pic/, 1,2-dinitrophenol, sodium tetraphenylborate and the $NH_4[Co/NH_3/_2 /NO_2/_4]$ complex.

Gold is selectively extracted by nitrobenzene from H_2SO_4 solutions /5 M - pH=6/ with high distribution coefficients as an $[Au/tu/_2^+ . Co/NH_3/_2 /NO_2/_4^-]$ complex. The dissociation constant of this complex in water-saturated nitrobenzene, calculated from conductivity measurements, is equal to $/2,14\pm0,15/.10^{-3}$. The extraction constants have been determined at different temperatures $/K_{ex} = 5,2\pm0,3$ at $20^\circ C$ and pH 5.05/ and thermodynamic parameters of the extraction have been calculated $/\Delta G = 5,5\pm0,05$ kkal/mol, $\Delta H = -0,9\pm0,05$ kkal/mol; $\Delta S = 15,2\pm0,2$ kal/mol.grad/. Gold can be selectively separated from silver after the extraction from the solutions of pH 4-7. Gold is also extracted completely from solutions of pH > 6 in the presence of picric acid.

Silver can be extracted with the highest distribution coefficients in the presence of picric acid or anionic cobalt complex. In the presence of picric acid the extracted silver complex has the composition $[Ag/tu/_3^+.Pic^-]$ /$pK_{ex}= 4,4\pm0,3$ at $20^\circ C$ and pH=3; $\Delta G=5,1\pm0,05$ kkal/mol; $\Delta H=-0,75\pm0,05$ kkal/mol; $\Delta S=14,6\pm0,2$ kal/mol.grad/. The dissociation constant of the complex in nitrobenzene is $/7,10\pm0,05/.10^{-7}$. Distribution coefficients of gold/I/ and silver complexes decrease when the metal concentrations are increased up to 8.10^{-5} and 5.10^{-4}M respectively.

The solutions of the $[Au/tu/_2^+ . Co/NH_3/_2/NO_2/_4^-]$ and $[Au/tu/_2^+.Pic^-]$ complexes were used as liquid membranes of $Au/tu/_2^+$ - selective electrodes. We used 2.10^{-3} M solution of $Au/tu/_2^+$ in 0,45 M Na_2SO_4 solution /pH=4,0/ as reference one. The measurements were made in 0,45 M Na_2SO_4 solutions as well. The electrode response is linear over a range of $10^{-2} - 10^{-5}$ g.ion/l /Fig.1., Table 1/. The detection limits of gold are equal to 1.10^{-5} or 1.10^{-4} M if the complexes with cobalt or picric acid are used respectively. The dependence of the $Au/tu/_2^+$ - selective electrode potential on the acid concentration is shown in Fig. 2. The potentials for the $Au/tu/_2^+$ - electrode

306

are constant in the solutions with the acidity from 4N H_2SO_4 to pH=6, the range corresponds to the distribution coefficients of gold in the extraction system.

Solutions of the $[Ag/tu/_3^+ \cdot Pic^-]$ complex in nitrobenzene as liquid membrane of the $Ag/tu/_3^+$ - selective electrode have been used. We used 2.10^{-3} M solution of $Ag/tu/_3^+$ in 0,45 M Na_2SO_4 with pH=3 as reference one. The measurements were made in 0,45 M Na_2SO_4 solutions containing 0,3 M thiourea. The electrode properties are shown in Table 1 and Fig. 3. The potential of the $Ag/tu/_3^+$ - electrode is constant in solutions with the acidity from 2 N H_2SO_4 to pH=4 /Fig.4/.

RESULTS AND DISCUSSION

The analysis of the electrochemical and extraction data leads to the conclusion that the properties of the systems are determined by the stability of the metal complexes; the properties of the anion-forming species and distribution coefficients of the ion association complexes.

The linear electrode response depends on the solvent and decreases in the order: chlorobenzene $<$ 1,2-dichloroethan $<$ nitrobenzene for all the systems studied. The distribution coefficients of the complexes increase in this order, too. As a rule, the detection limit for gold is lower than that for silver and the distribution coefficient of gold is higher in comparison with the distribution coefficient of silver in the some solvent. We confirmed that the detection limit depends on the ion-exchange concentration in the liquid membrane and decreases with ion-exchanger concentration. However, there is a lower limit of the ion-exchanger concentration which is connected with the increase of the ohmic resistance of the membrane and with the equilibrium attainment. The optimal concentration of the ion-exchanger in the membrane is from 5.10^{-3} to 1.10^{-3} M.

The concentration of potential determining ions depends on the hydrogen ion concentration and this is the explanation for the potential dependence of the electrochemical species on hydrogen ion concentration. The stability of the gold thiourea

complex $/\log\beta_2=25,0/2/$ is much higher than that of the silver
one $/\log\beta_2=13,6/3/$. As a result, the gold-electrode potential is constant within a wider pH-range in comparison with
the silver system. The lower stability constant of the silver
thiourea complex results in a wider pH range for silver electrode at a higher thiourea concentration /Fig. 4/. The potentials of the $Au/tu/_2^+$ - and $Ag/tu/_3^+$ - electrodes increase in
strong acid solutions. This can probably be explained by the
destruction of the ion-exchanger compounds and change in the
diffusion potential of external reference electrode - test
solution. It should be emphasized that the electrode function
slope for the $Ag/tu/_3^+$ - electrode and the distribution coefficient change at about the same silver concentration.

The potentiometric selectivity coefficients of the electrodes were calculated by the Srinivasan and Rechnitz method /4/.
The numerical values of the selectivity coefficients /Table 2./
depend on measurement conditions but in all cases the selectivity series is the same and for the $Au/tu/_2^+$ - electrode follows
the order: $Ag/I/> Cu/II/> Fe/III/> Fe/II/> Zn/II/, Ni/II/$.
The $Ag/tu/_3^+$ and $Au/tu/_2^+$ ions are the most interfering ions in
the determination of gold and silver, respectively.
There is a correlation between the free energy of the extraction and the selectivity coefficients. Figure 5 shows the
linear dependence of the selectivity coefficients upon the
free energy of the extraction for the $Au/tu/_2^+$-selective electrode in the case of $[Au/tu/_2^+ \cdot Co/NH_3/_2/NO_2/_4^-]$, as ion-exchanger. The interference of metal cations were estimated from the
stability constants of the corresponding metal-thiourea complexes. It is interesting to point to small changes of the
selectivity coefficients by substituting the membrane solvents.
If one changes chlorobenzene for nitrobenzene, the selectivity
coefficients change very little for the $Ag/tu/_3^+$-electrode but
much more for the $Au/tu/_2^+$-electrode. It can be explained by a
considerable difference in the dissociation constants for the
corresponding ion-exchangers.

SUMMARY

The liquid membrane ion-selective electrode can be used for the determination of $Au/tu/_2^+$ and $Ag/tu/_3^+$ ion concentration. The selectivity coefficients and other analytical parameters show the dependence on the properties of the ion-exchanger materials, solvents and the measurement conditions. The electrodes can be used both in research and industry.

REFERENCES

1. O.M.Petrukhin, Yu.V.Shavnja, A.S.Bobrova, Yu.M.Chikin. Journ.Inorg.Chem./Russian/ in press
2. B.N.Peshevickij, V.N.Belevancev, S.V.Zemskov. Izv. SO AN USSR, ser.chim.nauk, 4, N 2, p.29, /1976/
3. S.N.Khodaskar, D.D.Khandkar, Current Sci., 33, 339 /1964/
4. K.Srinivasan, G.A.Rechnitz, Anal.Chem., 41, 1203 /1969/

Table 1. Properties of M/tu_n^+-selective electrodes

Electrode	Ion-exchanger, $C = 1.10^{-3}M$	Solvent	E_{o1} mv	Slope	Detection limit
Au/tu_2^+ - SE	$Au/tu_2^+/Co/NH_3/_2/NO_2/_4^-$	Chlorobenzene	$118,7\pm0,7$ x/	$54,0\pm0,5$	$3,2\ 10^{-5}$
		1,2-Dichloro-ethane	$118,5\pm0,5$	$57,2\pm0,3$	$5,5\ 10^{-5}$
		Nitrobenzene	$120,3\pm0,9$	$58,0\pm0,8$	$1,0\ 10^{-5}$
	$Au/tu_2^+/C_6H_3N_3O_7^-$	Nitrobenzene	$119,5\pm0,9$ x/	$57,5\pm0,9$	$9,9\ 10^{-4}$
Ag/tu_3^+ - SE	$Ag/tu_3^+/C_6H_3N_3O_7^-$	Chlorobenzene	$127,1\pm0,8$ xx/	$54,2\pm1,1$	$2,8\ 10^{-4}$
		1,2-Dichloro-ethane	$128,0\pm1,1$	$54,4\pm1,2$	$1,2\ 10^{-4}$
		Nitrobenzene	$129,4\pm0,7$	$55,1\pm1,2$	$7,2\ 10^{-5}$

x/ Reference solution: $2.10^{-3}M$ Au/tu_2^+ in 0,45 M Na_2SO_4, pH=3,95

xx/ Reference solution: $2.10^{-3}M$ Ag/tu_3^+ in 0,45 M Na_2SO_4, pH=2,05

Table 2. The selectivity coefficients of the Au/tu/$_2^+$ - and Ag/tu/$_3^+$- electrodes. Concentration of the ion-exchanger [Au/tu/$_2^+$. Co/NH$_3$/$_2$./NO$_2$/$_4^-$] or [Ag/tu/$_3^+$. .Pic$^-$] 1.10^{-3} M in chlorobenzene /CB/ and nitrobenzene /NB/

Interfering ions	Au/tu/$_2^+$-SE CB	,pH 4,0 NB	Ag/tu/$_3^+$-SE CB	,pH 2,0 NB
Au/tu/$_2^+$	I	I	0,55	0,53
Ag/tu/$_3^+$	0,59	0,38	I	I
Cu/tu/$_4^+$	0,33	0,18	0,43	0,30
Fe/tu/$_2^{3+}$	0,02	0,008	0,05	0,045
Zn/tu/$_2^{2+}$	0,015	0,005	0,017	0,012
Pd/tu/$_3^{2+}$	$8,0\ 10^{-3}$	$1,0\ 10^{-4}$	$1,2\ 10^{-2}$	$9\ 10^{-3}$
Tu	$6,0\ 10^{-2}$	$4,0\ 10^{-2}$	$6,0\ 10^{-2}$	$5,3\ 10^{-2}$

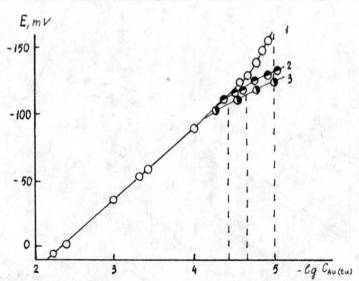

Figure 1. Effect of solvents on the response of the Au/tu/$_2$-electrode. The dotted lines show the detection limits.
Membrane: 1.10^{-3}M [Au/tu/$_2^+$.Co/NH$_3$/$_2$/NO$_2$/$_4^-$] in nitrobenzene/1/, 1,2-dichloroethane/2/, chlorobenzene/3/.
Test solution: [Au/tu/$_2$]$_2$SO$_4$ in 0,45 M Na$_2$SO$_4$,pH=3,95

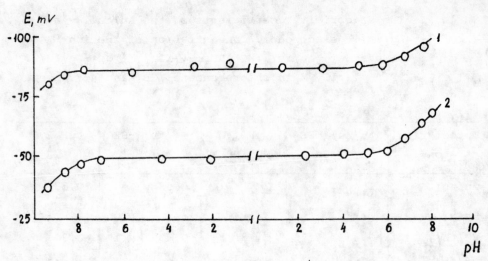

Figure 2. Potential change of the $Au/tu/_2^+$-electrode with the pH of solution.

Membrane: $1.10^{-3}M$ $[Au/tu/_2^+ \cdot Co/NH_3/_2/NO_2/_4^-]$ in nitro-benzene. Test solution: $[Au/tu/_2]_2SO_4$ in $0,45M$ Na_2SO_4; gold concentration is 1.10^{-4} /1/ or $5.10^{-4}M$ /2/

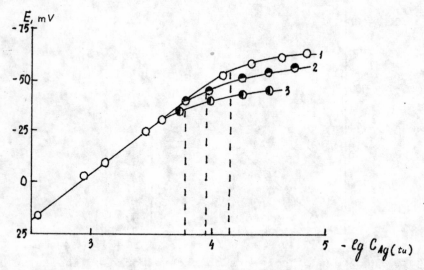

Figure 3. Effect of solvents on the response of the $Ag/tu/_3^+$-electrode. The dotted lines show the detection limits.

Membrane: $1.10^{-3}M$ $[Ag/tu/_3^+ \cdot Pic^-]$ in nitrobenzene/1/, 1,2-dichloroethane/2/, chlorobenzene/3/.

Test solution: $[Ag/tu/_3]_2$ SO_4 in $0,45$ M Na_2SO_4, pH=2,05

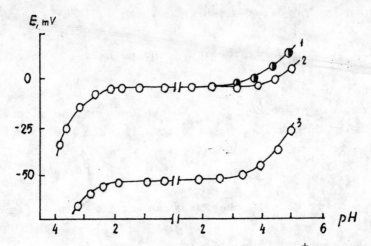

Figure 4. Changes of the potential of the $Ag/tu/_3^+$-electrode
as a function of pH.
Membrane: $1.10^{-3} M[Ag/tu/_3^+ . Pic^-]$ in nitrobenzene.
Test solution: $[Ag/tu/_3]_2 SO_4$ in 0,45 M $Na_2 SO_4$.
Silver concentration $1.10^{-3} /1,2/$ and $1.10^{-4} m /3/$,
thiourea concentration $0,1/1,3/$ and $0,3 M /2/$

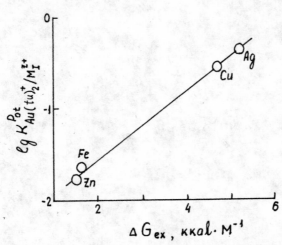

Figure 5. The dependence of the selectivity coefficients of
the $Au/tu/_2^+$-electrode on the free extraction energy
of the corresponding salts $[M/tu/_n^{Z+} . Co/NH_3/_2/NO_2/_4^-]_Z$.
Membrane: $1.10^{-3} M[Au/tu/_2^+ . Co/NH_3/_2/NO_2/_4^-]$ in nitrobenzene

313

QUESTIONS AND COMMENTS

Participants of the discussion: A.Hulanicki, O.M.Petrukhin

Question:
A diagram was presented on the dependence of the potential
on the sulphuric acid concentration. Why did the potential
decrease in more acidic solutions ?

Answer:
The reason for this is the formation of the complex in this
medium.

Question:
And the increase at higher pH ?

Answer:
I do not know exactly the reason for this.

MODEL CALCULATIONS ON THE STRUCTURE/SELECTIVITY RELATIONSHIP OF IONOPHORES

E. PRETSCH, M. GRATZL,* E. PUNGOR* AND W. SIMON

Swiss Federal Institute of Technology, Department of Organic Chemistry,
CH-8092 Zürich, Switzerland
*Institute for General and Analytical Chemistry, Technical University Budapest,
H-1521 Budapest, Hungary

ABSTRACT

A model has been developed for the calculation of the interaction energy of an ion with a ligand molecule. The interaction
energy is described as a sum of atom-ion interactions which
are represented by simple potential functions. The corresponding parameters are derived from ab initio calculations on
complexes of small model ligand molecules. Clear boundary conditions are given for the transferability of these pair potentials.

INTRODUCTION

Model calculations on ion-ligand interactions are in de-
mand for more efficient design of highly selective ionophores
for ion-selective liquid membrane electrodes. The selectivities of idealized neutral-carrier membrane electrodes are defined by the free energies of transfer of the corresponding
ions from the aqueous phase into the complexed form in the
membrane phase.

In the present contribution interaction energies between an
ion I and a ligand molecule L are calculated for the gas phase
as a sum of interactions of the ion with the individual atoms
A_i of the ligand molecule:

$$E_{IL} = \sum_i E_{IA_i} \tag{1}$$

The corresponding pair potentials E_{IA_i} are assumed to be a
simple function of the atom-ion distance (r), of the charge
of the atom (q_{A_i}), and of the charge of the ion (q_I):

$$E_{IA_i} = -A_i/r^6 + B_i/r^{12} + C_i q_I q_{A_i}/r \tag{2}$$

315

Atoms for which all α- and β-neighbours have the same atom number and the same hybridization are assumed to have the same set of parameters A_i, B_i, and C_i: they belong to the same class. This classification has proven satisfactory for all cases studied so far.

RESULTS

The interaction energies of Li^+ and Na^+ with the model ligands given in Fig. 1 were calculated for 649 respectively 617 complexes (cf. [1-3]). The parameters A_i, B_i, and C_i for the 28 classes involved were determined by fitting Eqs. (1) and (2) to ab initio data using the least-square approach [4] (see Table 1 and 2).

The ionophore 18-crown-6 consists of three classes of atoms. The hydrogen atoms are represented by class No. 8, the carbon atoms by class No. 20 (see Table 1 and 2). Although the oxygen atoms are not exactly represented by any of the classes derived so far, they can be approximated using class No. 26. The small difference between the values of the parameters of class No. 25 $(OC(H_3)C(H_3))$ and No. 26 $(OC(H_3)C(H_2C))$ indicates that class No. 26 should be a good approximation for an oxygen atom of the type $OC(H_2C)C(H_2C)$. Using these classes, the interaction of Na^+ with 18-crown-6 was calculated for three different conformations of the ligand molecule (see Fig. 2). There is only a very weak stabilizing interaction between Na^+ and 18-crown-6 in its conformation as found [5] for the free ligand (see Fig. 2 A). If the conformation determined for the K^+ complex [6] is assumed, the model directs the Na^+ ion to exactly the position where K^+ was found by X-ray cristallography (Fig. 2 B). For the conformation of 18-crown-6 in its Na^+ complex [7] the model again locates the Na^+ ion in exactly the same position as found experimentally. The strongest stabilizing interaction between Na^+ and the ligand was found for this conformation.

The conformation of an ionophore molecule may be drastically changed by complexation even for very simple ligands (cf. Table 3). For model calculations therefore both ion-ligand interaction energies and conformational energies should be considered. Work is in progress for including the conformational energy contributions using a semi-empirical approach (PCILO [8]).

REFERENCES

1. G. Corongiu, E. Clementi, E. Pretsch, and W. Simon, J. Chem. Phys. 70, 1266 (1979).
2. G. Corongiu, E. Clementi, E. Pretsch, and W. Simon, J. Chem. Phys. 72, 3096 (1980).

3. E. Pretsch, E. Clementi, G. Corongiu, A. Neszmélyi, and W. Simon, in preparation.
4. J. P. Chandler, QCPE. 307 (1975).
5. J. D. Dunitz and P. Seiler, Acta Cryst. B 30, 2739 (1974).
6. P. Seiler, M. Dobler, and J. D. Dunitz, Acta Cryst. B 30, 2744 (1974).
7. M. Dobler, J. D. Dunitz, and P. Seiler, Acta Cryst. B 30, 2741 (1974).
8. P. Claverie, J. P. Daudey, S. Diner, Cl. Giessner-Prettre, M. Gilbert, J. Langlet, J. P. Malvien, and U. Pincelli, QCPE. 220 (1974).

ACKNOWLEDGEMENT

This work was partly supported by the Swiss National Science Foundation.

Table 1. Parameters for the atomic pair potentials of Li^+ [1])

Atom	Group	No.	Code	A	B	C
H	CH_3	1	$HC(H_2C)$	$1.8 \cdot 10$	$7.5 \cdot 10^2$	0.87
		2	$HC(H_2C^*)$	$2.9 \cdot 10^{-2}$	$3.8 \cdot 10^3$	0.67
		3	$HC(H_2N)$	$7.2 \cdot 10$	$2.3 \cdot 10^3$	0.83
		4	$HC(H_2O)$	$2.8 \cdot 10^2$	$5.0 \cdot 10^3$	0.71
		5	$HC(H_2S)$	$7.2 \cdot 10$	$1.1 \cdot 10^3$	0.77
	CH_2	6	$HC(HCC)$	$9.6 \cdot 10$	$4.5 \cdot 10^3$	0.80
		7	$HC(HCN)$	$2.9 \cdot 10$	$1.7 \cdot 10^3$	0.46
		8	$HC(HCO)$	1.9	$7.0 \cdot 10^3$	0.65
		9	$HC(HCS)$	$2.6 \cdot 10^2$	8.1	0.92
C	CH_3	10	$CH_3C(H_2C)$	0.0	$3.7 \cdot 10^4$	0.92
		11	$CH_3C(H_2N)$	0.0	$5.7 \cdot 10^4$	1.0
		12	$CH_3C(H_2S)$	0.0	$2.0 \cdot 10^5$	1.1
		13	$CH_3C^*(=ON)$	$2.6 \cdot 10^3$	$3.7 \cdot 10^5$	0.89
		14	$CH_3N(CC^*)$	$3.9 \cdot 10^{-1}$	$7.3 \cdot 10^4$	1.1
		15	$CH_3O(C)$	0.0	$1.7 \cdot 10^5$	0.79
		16	$CH_3S(C)$	0.0	$3.6 \cdot 10^5$	0.75
	CH_2	17	$CH_2C(H_3)C(H_2O)$	$3.7 \cdot 10^2$	$7.0 \cdot 10^4$	0.93
		18	$CH_2C(H_3)N(CC^*)$	8.4	$3.6 \cdot 10^4$	0.53
		19	$CH_2C(H_2C)O(C)$	$1.6 \cdot 10^{-1}$	$1.7 \cdot 10^5$	0.15
		20	$CH_2C(H_2O)O(C)$	$9.4 \cdot 10^2$	$5.9 \cdot 10^5$	0.51
		21	$CH_2C(H_3)S(C)$	0.0	$2.5 \cdot 10^5$	0.43
C^*	$C=O$	22	$C^*C(H_3)N(CC)=O$	0.0	$3.7 \cdot 10^4$	1.5
N	$N-C=O$	23	$NC(H_3)C(H_3)C^*(=OC)$	$1.4 \cdot 10^{-1}$	$4.0 \cdot 10$	$6.2 \cdot 10^{-6}$
		24	$NC(H_3)C(H_2C)C^*(=OC)$	$1.4 \cdot 10^{-1}$	$4.0 \cdot 10$	$6.2 \cdot 10^{-6}$
O	$C-O-C$	25	$OC(H_3)C(H_3)$	$1.7 \cdot 10^2$	$5.0 \cdot 10^3$	0.60
		26	$OC(H_3)C(H_2C)$	$9.3 \cdot 10$	$2.8 \cdot 10^3$	0.66
O^*	$O=C$	27	$O^*=C(CN)$	0.0	$2.5 \cdot 10^3$	1.3
S	$C-S-C$	28	$SC(H_3)C(H_2C)$	$5.7 \cdot 10^2$	$1.2 \cdot 10^5$	0.0

[1]) To obtain the interaction energy E_{IL} in kcal/mol the dimension of the radii r is [Å], A and B are used as given in the Table, and C must be multiplied by a transformation factor of $627.503/1.889763 = 332.060$ to convert length from [Å] to a.u. and energies from a.u. to kcal/mol.

*) Indicates sp^2-hybridization.

Table 2. Parameters for the atomic pair potentials of Na^{+} [1]

Atom	Group	No.	Code	A	B	C
H	CH$_3$	1	HC(H$_2$C)	4.7	$3.5 \cdot 10^3$	0.93
		2	HC(H$_2$C*)	$4.5 \cdot 10$	$7.3 \cdot 10^3$	0.63
		3	HC(H$_2$N)	$2.7 \cdot 10^2$	$3.0 \cdot 10^4$	0.84
		4	HC(H$_2$O)	$2.6 \cdot 10^2$	$9.3 \cdot 10^3$	0.88
		5	HC(H$_2$S)	8.6	0.0	0.95
	CH$_2$	6	HC(HCC)	$1.4 \cdot 10$	$1.4 \cdot 10^4$	0.89
		7	HC(HCN)	$1.0 \cdot 10$	0.0	0.68
		8	HC(HCO)	0.0	$1.9 \cdot 10^5$	0.87
		9	HC(HCS)	$7.6 \cdot 10$	$2.1 \cdot 10^3$	0.94
C	CH$_3$	10	CH$_3$C(H$_2$C)	$1.1 \cdot 10^3$	$4.4 \cdot 10^5$	0.55
		11	CH$_3$C(H$_2$N)	0.0	$1.4 \cdot 10^5$	1.0
		12	CH$_3$C(H$_2$S)	0.0	$5.6 \cdot 10^5$	1.0
		13	CH$_3$C*(=ON)	5.0	$9.4 \cdot 10^4$	0.87
		14	CH$_3$N(CC*)	$7.8 \cdot 10^{-2}$	$3.3 \cdot 10^5$	1.1
		15	CH$_3$O(C)	$1.5 \cdot 10$	$4.3 \cdot 10^5$	1.1
		16	CH$_3$S(C)	0.0	$2.3 \cdot 10^5$	0.86
	CH$_2$	17	CH$_2$C(H$_3$)C(H$_2$O)	$2.7 \cdot 10^2$	$4.9 \cdot 10^4$	0.85
		18	CH$_2$C(H$_3$)N(CC*)	0.0	$2.3 \cdot 10^5$	0.93
		19	CH$_2$C(H$_2$C)O(C)	0.0	$4.6 \cdot 10^5$	1.3
		20	CH$_2$C(H$_2$O)O(C)	$2.4 \cdot 10^3$	$2.0 \cdot 10^6$	1.1
		21	CH$_2$C(H$_3$)S(C)	0.0	$1.8 \cdot 10^5$	0.61
C*	C=O	22	C*C(H$_3$)N(CC)=O	$2.2 \cdot 10^2$	$2.9 \cdot 10^4$	1.6
N	N-C=O	23	NC(H$_3$)C(H$_3$)C*(=OC)	0.0	$3.5 \cdot 10^4$	0.0
		24	NC(H$_3$)C(H$_2$C)C*(=OC)	0.0	$3.5 \cdot 10^4$	0.0
O	O-C-O	25	OC(H$_3$)C(H$_3$)	7.9	$2.2 \cdot 10^4$	0.55
		26	OC(H$_3$)C(H$_2$C)	$1.0 \cdot 10^2$	$2.3 \cdot 10^4$	0.56
O*	O=C	27	O*=C(CN)	2.5	$2.2 \cdot 10^3$	1.4
S	C-S-C	28	SC(H$_3$)C(H$_2$C)	$1.6 \cdot 10^3$	$4.6 \cdot 10^5$	$1.1 \cdot 10^{-3}$

[1] To obtain the interaction energy E_{IL} in kcal/mol the dimension of the radii r is [Å], A and B are used as given in the Table, and C must be multiplied by a transformation factor of 627.503/1.889763=332.060 to convert length from [Å] to a.u. and energies from a.u. to kcal/mol.

*) Indicates sp^2-hybridization.

Table 3. Ion-ligand interaction energy (E_{IL} in kcal/mol) and conformational energy (E_{conf} in kcal/mol) for the complexes of Li^+ and Na^+ with 1,2-dimethoxy ethane in different conformations.

Ion	Dihedral angle OCCO	Ab initio calculations			Model calculations
		E_{conf}	E_{IL}	$E_{IL}+E_{conf}$	E_{IL}
Li^+	$0°$	7.2	-58.3	-51.1	-58.2
	$30°$	4.6	-56.4	-51.8	-55.8
	$60°$	1.3	-49.0	-47.7	-47.3
	$180°$	0.0	-28.7	-28.7	-27.5
Na^+	$0°$	7.2	-41.1	-33.9	-40.6
	$30°$	4.6	-40.1	-35.5	-39.6
	$60°$	1.3	-35.9	-34.6	-35.8
	$180°$	0.0	-19.0	-19.0	-15.6

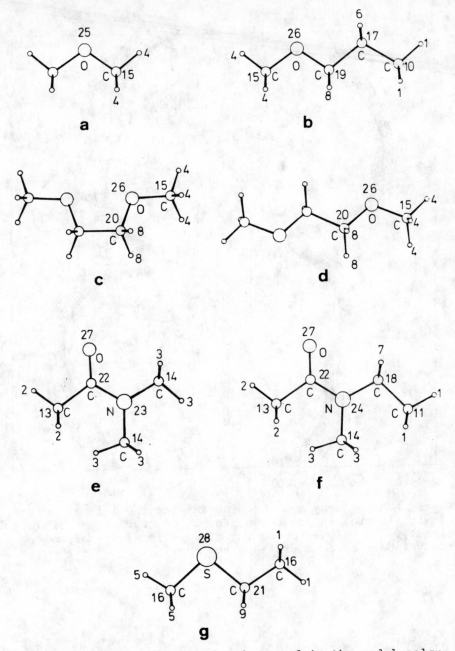

Figure 1. Structure of the molecules used in the model calculations. a: dimethyl ether; b: methyl propyl ether; c and d: 1,2-dimethoxy ethane; e: N,N-dimethylacetamide; f: N-ethyl-N-methylacetamide; g: ethylmethyl sulfide. The numbers of the individual atoms designate the corresponding classes (cf. Table 1 and 2).

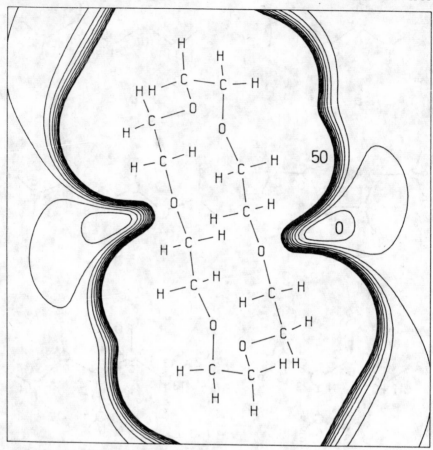

Figure 2A. Isoenergy contour diagrams for the interaction of Na$^+$ with 18-crown-6 for the conformation of the free ligand molecule. The calculated energy minimum is -4.4 kcal/mol. The interval between contours is 5 kcal/mol.

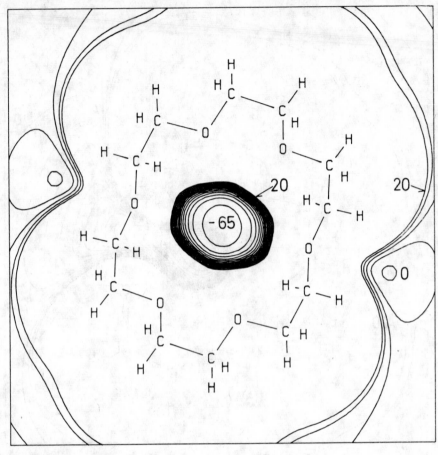

X= -6.00 Y= 6.00 Z= 0.00   X= 6.00 Y= 6.00 Z= 0.00

X= -6.00 Y= -6.00 Z= 0.00

Figure 2B. Isoenergy contour diagrams for the interaction of Na[+] with 18-crown-6 for the conformation of the K[+]-complex. The calculated energy minimum is -66.8 kcal/mol. The interval between contours is 5 kcal/mol.

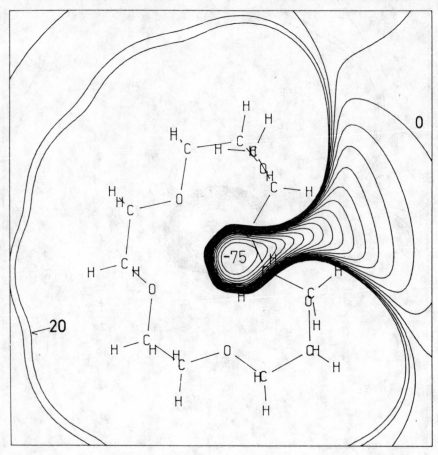

Figure 2C. Isoenergy contour diagrams for the interaction of Na$^+$ with 18-crown-6 for the conformation of the Na$^+$-complex. The calculated energy minimum is -80.8 kcal/mol. The interval between contours is 5 kcal/mol.

QUESTIONS AND COMMENTS

Participants of the discussion: A.Lewenstam, Y.Umezawa,
E.Pretsch

Question:
Some energy values mentioned in the paper seem to be arbitrary.
How were the radii of the species chosen ? Did you try to
treat the problem based on the extended Hückel theory ?

Answer:
The function shown contained the term r, but it is not the
radius of the species but the distance between the atom,
ligand and ion. We did not do any calculation with the extended Hückel equation, but many other workers did. These
results were wrong with respect to the geometry and the
energy of interaction. Many semi-empirical calculations were
tried but the results were poor.

Question:
Could your theory be applied to predict new ligands with
higher selectivities ?

Answer:
Of course this is what we should like to do. The idea would
be: you tell the computer which ion you would like to
determine and the computer tells you the appropriate ligand.
But we would be happy with the opposite: we propose ligand
structures and calculate the interaction energies at least.
At present we are trying this for known structures and learn-
ing how they fit. The next step will be of course to try to
design ligands and calculate the selectivities.

METHOD FOR CALCULATION OF THE CONCENTRATION DEPENDENCE OF k^{Pot} VALUE FOR LIQUID MEMBRANES WITH CHARGED CARRIER

J. SENKYR AND J. PETR

Analytical Chemistry Department, J. E. Purkyne University, 61137 Brno, Czechoslovakia

ABSTRACT

A method has been developed for the calculation of k^{Pot} at any concentration combination of the measured and the interfering ions valid for liquid membranes with charged carrier. A time dependent parameter has been proposed for the characterizing of the critical concentration range of liquid membranes where the original characteristics suddenly changes.

INTRODUCTION

The selectivity of ion sensitive electrodes is characterized by the selectivity coefficient k^{Pot}, the value of which is a variable. It is commonly known, that the k^{Pot} value changes with changing concentrations of the measured (c_A) and also of the interfering (c_X) ions. Figure 1 shows the concentration plot of selectivity coefficients for several anions valid for the nitrate electrode with liquid membrane. The decrease in c_X means only very small changes of k^{Pot} in the range of higher c_X values. But since a certain concentration value (the importance of this value will be later demonstrated), the further concentration decrease causes a sudden change in selectivity coefficient, the value of which limits to 1; $\log k^{Pot}$ is zero in all cases. It means that the ion selective electrode is no more selective in the range of low concentrations of interfering ions.

22*

Variations in k^{Pot} values are commonly explained by changes in concentration of the measured ion and also of the interfering ion in the vicinity of the membrane surface /1-3/. The concentration changes are due to ion exchange between the aqueous (measured) solution and the organic membrane solution. This ion exchange is schematically shown in Figure 2. The original bulk concentrations c_A and c_X are used for the calculation of k^{Pot} values, meanwhile the potential value $\Delta\varphi$ is given by the real concentrations $c_A^\ast$ and $c_X^\ast$ of ions in the boundary layer.

Replacing the values c_A and c_X by the real concentrations in the boundary layer $c_A^\ast$ and $c_X^\ast$ in the Nikolsky equation (1) we get the real selectivity constant $B_{A,X}$ (according to Eisenman), which is independent on concentrations c_A and c_X:

$$E = E_0 \pm S \log (c_A + k_{A,X}^{Pot} c_X) \qquad (1)$$

$$c_A + k_{A,X}^{Pot} c_X = c_A^\ast + B_{A,X} c_X^\ast \qquad (2)$$

where S is the slope of the potential-response curve. Yoshida and Ishibashi /3/ rearranged Eq. (2) into the form of (3):

$$k_{A,X}^{Pot} = 1 + (B_{A,X} - 1) \frac{c_X^\ast}{c_X} \qquad (3)$$

The mathematical description of the electrode function requires the real concentrations of ions in the boundary layer.

THEORETICAL CONSIDERATIONS AND RESULTS

We derived relations for $c_A^\ast$ and $c_X^\ast$ valid for liquid membranes with charged carriers under five limiting assumptions:
1. total dissociation of the charged carrier in membrane
2. rapid ion exchange
3. negligibly small extraction of the charged carrier into the measured water solution
4. constant stirring of the measured solution
5. only univalent ions participate in the potential forming process.

Equilibration in ion exchange and mass transfer due to concentration gradients resulting in ion exchange were considered in deriving the equations for $c_A^\ast$ and $c_X^\ast$ (4 and 5):

$$c_A^\ast = c_A + \frac{U_X}{U_A}(c_X - c_X^\ast) \qquad (4)$$

$$c_X^\ast = c_X - \frac{c_A + B_{A,X}(c_X + Q) + \sqrt{\left[c_A + B_{A,X}(c_X - Q)\right]^2 + 4B_{A,X}Q\left(c_A + \frac{U_X}{U_A}c_X\right)}}{2\left(B_{A,X} - \frac{U_X}{U_A}\right)} \qquad (5)$$

with

$$B_{A,X} = \frac{\overline{U}_X}{\overline{U}_A} K_{exch} \qquad (6)$$

and

$$Q = \overline{U}_A U_A^{-1} U_X^{-1/2} (4\pi RT)^{-1/2} t^{-1/2} d\overline{c}_J \qquad (7)$$

where
U_A, U_X ... mobilities of A and X in water
$\overline{U}_A, \overline{U}_X$ ... mobilities of A and X in membrane
d ... thickness of Nernstian layer
$\overline{c}_J$ ... charged carrier concentration in membrane
t ... time
K_{exch} ... ion exchange constant

$B_{A,X}$ means the already mentioned selectivity constant according to Eisenman. Parameter Q depends mainly on charged carrier concentration in the membrane $\overline{c}_J$ and on time t. Comparable Q values are obtained measuring the electrode potential in a constant time after electrode immersion into the measured solution - for instance after 100 seconds. The Q_{100} values for different anions do not differ much:

Anion	ClO_4^-	SCN^-	I^-	Br^-	Cl^-
$-\log Q_{100}$	5,87	6,02	6,08	5,98	5,93

329

The membrane with Crystal Violet nitrate in nitrobenzene, $\bar{c}_J = 10^{-4} M$, was used.

The Q value is identical with that concentration c_X, in which the maximal change in k^{Pot} value is observed. In Fig. 3 the k^{Pot} versus c_X plots were registred at different $\bar{c}_J$. The lower $\bar{c}_J$, the lower Q_{100} values were determined. The experimental points are in excellent agreement with theoretically calculated curves. The Equation (8)

$$k_{A,X}^{Pot} = B_{A,X} - \frac{1}{2}\left[\frac{c_A}{c_X} + B_{A,X}(1 + \frac{Q}{c_X}) \right.$$

$$\left. - \sqrt{\left[\frac{c_A}{c_X} + B_{A,X}(1 - \frac{Q}{c_X})\right]^2 + 4B_{A,X}\frac{Q}{c_X}(\frac{c_A}{c_X} + \frac{U_X}{U_A})} \right] \quad (8)$$

used for calculation of the theoretical curves was obtained by combination of Eqs. (3) and (5). With the knowledge of $c_A^{\ast}$ and $c_X^{\ast}$ the value of k^{Pot} can be calculated for combination of any c_A and c_X.

DISCUSSION

When $c_X = Q$ ($c_A = 0$ and $U_X = U_A$) the Equation (8) is rearranged into the (9) one:

$$\log k_{A,X}^{Pot} = \frac{1}{2}\log B_{A,X} \quad (9)$$

As may be seen in Fig. 4 the Q value is a half wave concentration in plot $\log k^{Pot}$ vs. $\log c_X$. It is identical with the critical concentration of interfering ions, at which the maximal change in k^{Pot} values is observed.

Application of $c_A^{\ast}$ and $c_X^{\ast}$ in the Nikolsky equation /the combination of Equations (1), (2), (4) and (5)/ gives the electrode function (10). As Q is time dependent, also the potential E is a function of time (Fig. 5). The experimental time

$$E = E_o \pm S \log \frac{c_A + B_{A,X}(c_X - Q) + \sqrt{\left[c_A + B_{A,X}(c_X - Q)\right]^2 + 4QB_{A,X}(c_A + \frac{U_X}{U_A}c_X)}}{2} \tag{10}$$

dependence of the nitrate electrode potential, measured in solutions with nitrate and perchlorate ions, shows good agreement with the theoretically calculated curves. The time necessary for potential establishment in solutions with different composition can be calculated using the Equations (10) and (7). Analysing this relation, such regions of concentrations c_A and c_X were found, where no constant potential can be reached in practice. That are the cases, when c_X is close to the Q value and c_A is close to the product $QB_{A,X}$. The knowledge of the Q function is very important for potential measurement in solutions with interfering ions. We propose the Q value measured after 100 seconds, Q_{100}, as one of the important parameters characterising the electrode.

The Q value is useful also for the estimation of the error in determination of c_A. The relative error of nitrate determination caused by the presence of chlorides is demonstrated in Figure 6. The concentration ratio $c_{Cl}/c_{NO_3} = 10$ is constant for the whole range of the calibration plot. The Q_{100} value of the ORION 92-07 electrode is about 10^{-2} — 10^{-3}M, therefore the error even for relatively high NO_3^- concentration as 10^{-3}M is no more negligible. Our membranes with charged carrier concentration $\bar{c}_J = 10^{-4}$M have $Q_{100} \sim 10^{-6}$M and the relative error of 4% was found even for $c_A = 10^{-6}$M NO_3^- /4,5/.

The value Q_{100} for any liquid membrane can be easily determined by experiment. The calibration curve is measured for solutions containing only the interfering ion, for which $B_{A,X} > 1$. The potential registered 100 sec after immersion into the solution is plotted versus $\log c_X$ (Fig. 7). $c_X = Q_{100}$ is valid for that point in the calibration curve, for which $\Delta E/\Delta \log c_X$ is maximal.

The Q_{100} values depend in low extent also on the stirring velocity. It is hardly possible to keep constant thickness of the Nernstian layer (d in Fig. 2) in routine measurements,

therefore the approximate Q_{100} values can be determined only.
Nevertheless these Q_{100} values are very useful for the estimation of critical concentration range, in which the electrode
suddenly loses its selectivity and in which the potential
establishment is slow or even impossible.

REFERENCES

1. D.Midgley, Anal. Chem. 49, 1211 (1977)
2. A.Hulanicki and A.Lewenstam, Talanta 24, 171 (1977)
3. N.Yoshida and N.Ishibashi, Bull.Chem.Soc.Japan 50, 3189 (1977)
4. J.Šenkýř and J.Petr, Chem.Listy 73, 1097 (1979)
5. J.Šenkýř and J.Petr, in E.Pungor (Editor) "Ion-Selective Electrodes", Conference held at Budapest, Akadémiai Kiadó, Budapest 1977, p. 559.

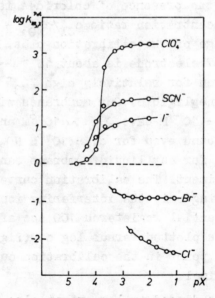

Figure 1. Dependence of selectivity coefficients k^{Pot} on c_X.
Measured with nitrate electrode $\bar{c}_J = 10^{-2}$M Crystal Violet nitrate ($CVNO_3$) in nitrobenzene.

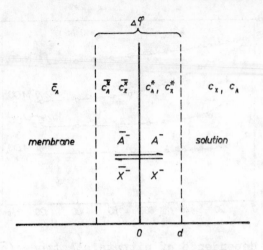

Figure 2. Schematic view of the ion exchange in the boundary layers.

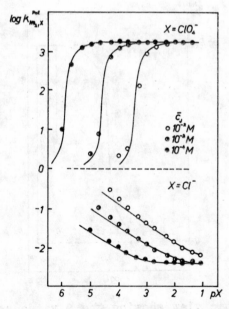

Figure 3. Dependence of selectivity coefficients $k^{Pot}_{NO_3, ClO_4}$
and $k^{Pot}_{NO_3, Cl}$ for membranes with different $\overline{c}_J$.

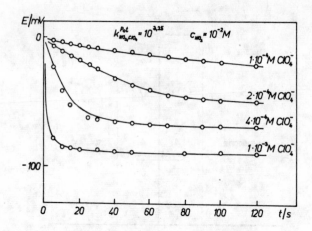

Figure 4. Time dependence of nitrate electrode ($\overline{c}_J = 10^{-3}$M CVNO$_3$) potential in nitrate solution $c_A = 10^{-2}$M with different content of perchlorate.

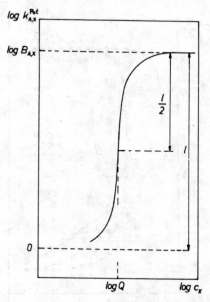

Figure 5. Schematic view showing the Q value as half wave concentration in the plot k^{Pot} vs. log c_X.

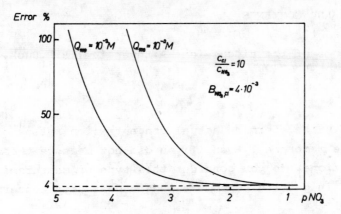

Figure 6. Influence of the parameter Q_{100} on relative error in nitrate determination. Calculated for solutions with $c_{Cl}/c_{NO_3} = 10$.

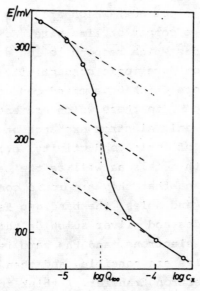

Figure 7. Determination of Q value from the experimental calibration curve. $\bar{c}_J = 10^{-3}$M CVNO$_3$, $c_A = 0$, X = ClO$_4^-$.

QUESTIONS AND COMMENTS

Participants of the discussion: A.Lewenstam, R.P.Buck,
J.Senkyr

Question:
Did you try to determine actual concentration ratios of ions
inside the membrane ? Even if you did not measure these
concentrations, do you consider this type of measurement to
be useful in getting some more independent data concerning
this model ?

Answer:
We have not yet measured the concentrations inside the
membrane, only calculated them.

Comment:
The subject was approached by Dr.Koryta by stating that we
should make a distinction between ion-selective electrodes
which operate on ionic conductivity exclusively from those
electron-exchanging electrodes which are also selective for
metal ions, such as a copper or zinc wire. Now the problem
is that the electrodes which respond to ion activities span
the entire range from pure ionic conductivity and pure ion
exchange through those which have mixed conductivity - both
ionic and electronic - to those which are entirely electronic
conductors and have only electron exchange at their surface.
In the extreme case of ionic conductivity, e.g. the glass
electrodes doped with metals as well as the LaF_3 electrode,
I think we would agree that they are ionic conductors, the
number of electrons and holes, the band gap is large. In the
case of silver halides and silver sulphide the band gaps are
getting smaller for electrons, and the same is true for ions,
so both conductivities are possible, and then you have the
metals with pure electron transfer. I think the hard problem
in making the cut-off is that in the middle you have the
halides where, on the one hand the electrode works because
there is silver-ion exchange at the surface, but there is
still an argument that there is enough free metal, so it can

336

work as a silver metal electrode. Furthermore, there exist silver halide based electrodes with solid inner contact which, at the membrane-electrolyte interface exchange silver ions, but at the silver wire-membrane interface an electron exchange must occur. This type of electrodes could not be considered, according to the conductivity mechanism, as ion-selective electrodes. However, we do not know exactly their conductivity mechanism. In my opinion, among ion-selective electrodes there is a branch of pure ionic, and a branch of pure electronic conductivity ones, and the branch between them, but all belong to the category of ion-selective electrodes.

COMPARATIVE STUDY OF THE STRUCTURE AND THE FUNCTION OF SOME COMMERCIAL COPPER-SELECTIVE ELECTRODES

J. SIEMROTH AND I. HENNIG

Sektion Chemie, Martin-Luther-Universität, 402-Halle, GDR

ABSTRACT

The sensing material of some commercial copper-selective electrodes have been studied by chemical analysis, reflected light microscopy, scanning electron microscopy and x-ray powder diffractometry. The results of our investigations are in contrast to the widespread opinion on the composition of such materials. Especially in materials made from a mixture of copper sulfide and silver sulfide we couldn't find free copper sulfide neither in form of CuS nor as $Cu_{2-x}S$. Only ternary phases could be identified besides an excess of silver sulfide. Furthermore, the electrochemical properties of these electrodes were investigated. But these results we don't want to report here.

INTRODUCTION

Till now a great variety of copper-selective electrodes has been described. A number of them is made by various producers and is commercially available. Comparative studies involving such commercially available electrodes are known (1),(2),(3), (4),(5), but are related to their electrochemical properties and their application for analytical purposes exclusively. Contrary to that, information on the composition of the sensing material and the phase assemblage therein is quite doubtful. Usually in publications the statements of the producers are taken over only. But these are quite often unsatisfactory.

The aim of our work was primary to determine the precise
composition of the sensing material of some commercial
copper-selective electrodes and to identify the phase
assemblage of such materials. Furthermore, the electrochemical
properties of these electrodes were investigated and put into
relation to the nature of the sensing material applied. Since
the results of our studies of the composition on the sensing
material seem to be more important, we will report here only
the first part of this program.

EXPERIMENTAL

Apparatus.

The copper-selective electrodes investigated are listed in
Table 1.

The phase assemblage of the sensing material of these electrodes were identified by using a Neophot II light microscope
(VEB Carl Zeiss), a scanning electron microscope JSM-U3 (JEOL),
and anX-ray diffractometer (Philips Micro 1011).

Procedure.

After finishing of the electrochemical measurements not reported here, the copper-selective electrodes were dismounted
with the exception of the Orion and Radiometer F 1112
electrodes. We got only on loan these two electrodes. The
sensing material of all dismounted electrodes were divided
into two parts. One part was used for the investigation with
the reflected light microscope and the scanning electron
microscope. The other part was powdered in an agate mortar.
From the powder of the material x-ray diffratograms were
recorded and finally the composition was determined by wet
chemical analysis.

RESULTS AND DISCUSSION

The results of our investigations on the composition of the
sensing material of some commercial copper-selective electrodes as well as the phase assemblage therein are summarized in

Table 1. It may be taken from it that the sensing materials made from a mixture of copper sulfide and silver sulfide do not contain free copper sulfide neither in form of CuS nor as $Cu_{2-x}S$. Only the ternary phases jalpaite and mckinstryite of formally univalent copper besides an excess of silver sulfide in form of the low temperature modification, acanthite, could be identified. This result is in contrast to the widespread opinion on the composition of the sensing material of such electrodes. But that means that also the theoretical approaches of the mechanism of the potential building process should be thought over. The presence of silver sulfide in excess in the sensing material is most likely due to the production technology. The same applies to the system copper selenide/silver selenide, where excess silver selenide were found. Possibly only in presence of excess silver sulfide or silver selenide relatively non-porous materials are obtained, due to the plasticity of these compounds. Nevertheless, in all materials made by pressing of powders small pores were found (Fig. 1-4). These pores may be the reason of a more or less pronounced memory effect. But these problems we don't want to discuss in detail.

A special case is the Radiometer Selectrode. Here, the electrode is activated by rubbing in a powdered sensing material into the surface of the electrode. This powdered sensing material according to our studies consists of a mixture of mckinstryite and acanthite. But from the results of the chemical analysis the phase assemblage can't be calculated. Probably, the powder also contains x-ray amorphous water soluble copper compounds. These may be formed by oxydation of copper sulfide or the ternary compound mckinstryite. This is also consistent with the time dependent extension of the linear range of the calibration curve.

In case of the Crytur and Radiometer F 1112 electrodes, according to the producers' instruction, the sensing material consists of a single crystal of the composition $Cu_{1,8}Se$ (berzelianite). In the sensing material of electrodes which had been used for extented periods, we were able to prove unambiguously the presence of umangite. It can't be distinguished from these data, whether the umangite is present in the material originally or whether it has been formed by reaction of berzelianite with aqueous solutions. The fact that the umangite is found only at the surface of the material (Fig.5) points to the latter possibility. Simultaneously, always whereumangite is present, small micro cracks are to be seen. These cracks may be due to a volume increase connected with the conversion of berzelianite into umangite. But when in case of berzelianite containing sensing materials contact with aqueous solutions a conversion into umangite really takes place, instability of the electrode function and finally decomposition of the material by crack formation should be expected.

Summarizing the results of our studies we can say that in no case single-phase materials could be found. Therefore, we also have to pose the question whether the term "homogeneous membrans" for such materials is false or at least irrelevant.

ACKNOWLEDGEMENT

We would like to thank Mr. Schumann, Institut für Festkörperphysik und Elektronenmikroskopie, Akademie der Wissenschaften, Halle/Saale for his help by taking the scanning electron photomicrographs. Thanks are also due to Prof. K. Burger, Institute for General and Analytical Chemistry, L. Eötvös University, Budapest, for the gift of an Orion copper-selctive electrode.

REFERENCES

1. E.H.Hansen, C.G.Lamm and J.Ruzicka, Anal.Chim.Acta $\underline{59}$, 403 (1972)
2. P. Lanza, ibid $\underline{105}$, 53 (1979)
3. D. Midgley, ibid $\underline{87}$, 7 (1976)
4. D. Midgley, ibid $\underline{87}$, 19 (1976)
5. J.C.Westall, F.M.M. Morel and D.N.Hume, Anal.Chem.$\underline{51}$,1792 (1979)

Table 1. Sensing Material of Some Commercial Copper-Selective Electrodes

Electrode	Composition (wt.%)	Phase Assemblage	Calculated Phase Composition (wt.%)
Orion 94-29		Jalpaite + Acanthite $Ag_{1,55}Cu_{0,45}S + Ag_2S$	
Metrohm EA 306-Cu	81,08% Ag; 5,22 % Cu	Jalpaite + Acanthite	42 % Jp + 58 % Ac
Crytur 29-17	57,41 % Cu	Berzelianite + Umangite $Cu_{1,8}Se + Cu_3Se_2$	61 % Bz + 39 % Um
Radiometer F 1112 Cu		Berzelianite + Umangite	
Radiometer F 3002	42,65 % Ag; 24,95 % Cu	Mckinstryite + Acanthite $Ag_{1,2}Cu_{0,8}S + Ag_2S$	
Hermsdorf 4199.4	53,30 % Ag; 15,45 % Cu	Eucairite + Naumannite $AgCuSe + Ag_2Se$	61 % Eu + 37 % Nau

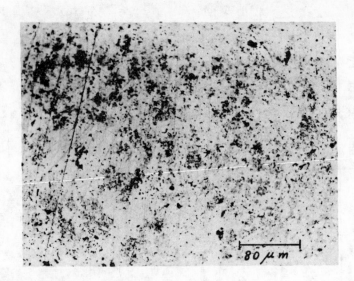

Fig.1. - Photomicrograph of the surface of an Orion copper-
selective electrode. Scale as indicated. Pores (black) in
jalpaite (gray) and acanthite (white).

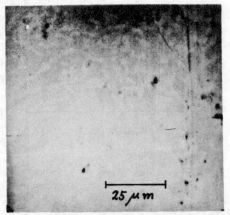

Fig.2. - Scanning electron photomicrograph of the structure
of a Metrohm copper-selective electrode. Scale as indicated.
Pores (black) in jalpaite (gray) and acanthite (white).

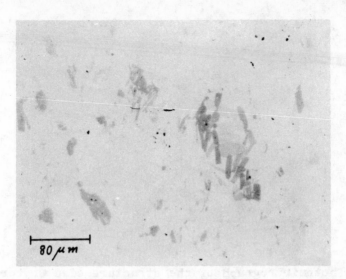

Fig.3. - Photomicrograph of the structure of a Hermsdorf copper-selective electrode. Scale as indicated. Pores (black) mainly in eucairite (white). Light to dark gray

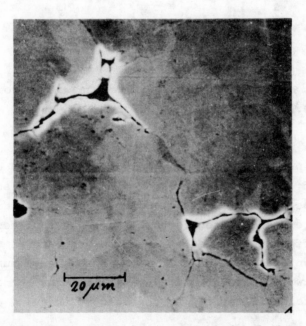

Fig.4. - Scanning electron photomicrograph of the structure of a Hermsdorf copper-selective electrode. Scale as indicated. Pores and micro cracks are clearly to be 'seen.

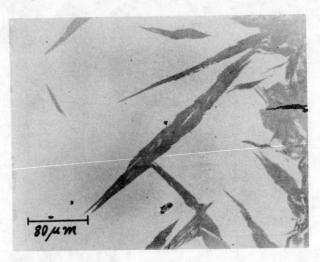

Fig.5. - Photomircrograph of the structure of a Crytur
copper-selective electrode. Scale as indicated.
Berzelianite (white) is replaced by umangite (gray). Note
the crack (black) within the umangite at the surface of
the material.

QUESTIONS AND COMMENTS

Participants of the discussion: A.Lewenstam, E.Pungor,
W.Simon, R.P.Buck, G.Johansson, J.Koryta

Comment:
The picture put at the end of the lecture seemed to be too
dramatic. If, for example we think of a platinum electrode,
it can not be considered as homogeneous, even if it is pure,
because there are different crystallographic surfaces which
may have different energies of electron transfer. I think
if we have an electrode made of a mixed substance such as a
mixture of two sulphides, we may consider it as homogeneous
even if different sites can be observed by scanning electron
microscopy.

Answer:
This is again a question of definition. But from the point of

346

view of the working mechanism of the electrode described, only the presence of a ternary compound is important.

Question:
How were the phase components of the surface determined ?

Answer:
The active material was studied by X-ray diffraction technique, light microscopy and electron microscopy, and all three gave identical results. Even with a light microscope one can see the different phases at the surface.

Question:
It was mentioned that only the ternary compound was responsible for the electrode function. We have made a number of good electrodes containing copper/II/ sulphide only.

Answer:
I have not investigated electrodes consisting of pure copper /II/ sulphide. In these studies I was concerned with commercially available mixed precipitate based electrodes and my statement about the ternary compound refers to these.

Question:
It was quite shocking to see that holes and cavities with diameters of up to ten μm occur at the electrode surface, in which solution may be trapped. These are sites with the lowest resistance and highest activity. Is not the response very slow as the solution has to be displaced from the pores before the electrode can respond to a new solution ?

Answer:
It is impossible to investigate the response on these spots only.

Question:
Had the electrodes internal metal contacts ? How can you ignore the equilibrium at the surface between copper exchange,

silver exchange and sulphide exchange ? Is it justified to say that silver sulphide is not important in the mechanism of the electrode response, if all the species are in equilibrium at the surface ? They must all be important as being in contact with the metal, what makes all of them active.

Answer:
Three kinds of the electrodes studied had metal contacts.

If we dip an electrode prepared from silver sulphide into a solution of copper/II/ ions, it will not respond to copper ions. However, a copper sulphide electrode as well as a ternary compound respond to copper/II/ ions.

Question:
We have made microscopic studies on a number of electrodes containing sulphides and never observed holes like the ones shown in the slides, and this seems to be very unusual.

Have any measurements been made to find out whether the different faces of the crystal have different activities ?

Answer:
Measurements of this type are is progress.

Comment:
Such measurements have already been made but it is very difficult to block the different faces of a single crystal electrode, as the blockig particles may migrate to the face which is to be studied.

Comment:
A paper by Rumanian authors is worth mentioning here, in which beautiful titration curves are shown for the titration of copper/II/ with EDTA in the presence of a silver sulphide electrode.

Answer:
It is quite obvious, as the silver sulphide electrode is sensitive to EDTA.

A MICRO-IMMUNOELECTRODE

Y. UMEZAWA, K. SHIBA, T. WATANABE, S. OGAWA AND S. FUJIWARA

Department of Chemistry, Faculty of Science, The University of Tokyo, Hongo, Tokyo 113, Japan

ABSTRACT

A new micro immunoelectrode system is reported, where thin-layer potentiometric measurement of complement and antibody levels in microliter serum was performed using tetrapentylammonium ion(TPA^+) loaded liposomes and TPA^+ ion-selective electrode.

INTRODUCTION

The present paper describes a novel method for studying antigen-antibody reaction and activation of complement on liposomes(1,2). The method involves a thin-layer micro-ion-selective electrode(3,4) which monitors the complement-mediated immune lysis of sensitized phospholipid liposomes in micro-liter solution. The liposomes are loaded with a concentrated solution of water soluble membrane-impermeable tetrapentyl-ammonium(TPA^+) ion. The TPA^+ ion retained within the liposomes is insensitive to a TPA^+ ion-selective electrode(ISE). Complement mediated lysis releases the TPA^+ ion into a dilute solution, where the ISE for the TPA^+ ion detects the amount of ions(Fig. 1). The response time of the TPA^+ ISE is sufficiently rapid($t_{95} < 30$ s) so that under the conditions of assays, the rate of the potential change is a valid measure of the rate at which the TPA^+ marker is released from the liposomes.

Complement-mediated release of marker species from phospho-lipid liposomes has been reported (5-8). However, none of the techniques previously used combines all the advantages of the present thin-layer micro-ion-electrode has. Potentiometric response from the ISE being proportional to the extent of TPA^+ ion release, optically opaque or clear 10 μL samples containing both immunologically lysed and nonlysed sacs may be continuously monitored at a given temperature(Fig. 1). Unlike glucose and similar molecules, which can be detected by their enzyme-mediated reactions(5), TPA^+ ions are not present in the biological materials used, and therefore, prior purification of these is not required. The use of a spin-label marker results in the increase in sensitivity of such assays(8). However, at the moment, routine application of esr is obviously limited by the expense of the spectrometer.

EXPERIMENTAL

The multilamellar liposomes were prepared as before(6) using dipalmitoylphosphatidylcholine, cholesterol, dicetyl-phosphate and an appropriate antigen in molar ratio of 2:1.5: 0.2:0.1. The dried lipids were swollen in 0.15 M TPA$^+$ aqueous solution. Untrapped marker ions, TPA$^+$ ions, were removed by dialysis against a modified Tris buffer saline(2) for ca.4 hours or by centrifugation(15,500xg) for 15 min. each at 0°C with five changes of volume of the modified Tris buffer saline. The TPA$^+$ ISE was made following the modified version of the Higuchi's method(2). A plate-shaped Ag/AgCl reference electrode was made by anodic oxidation of silver plate(4x7 cm, 2mm thick) in 0.1 M KCl solution at +0.5 V vs. SCE for 5 min. Although the Ag/AgCl plate reference electrode is in direct contact with the sample solution, the potential shift of this electrode against Cl$^-$ interference from one serum sample to another was found negligible. Also, other constituents of sera, if any, particularly proteins, did not cause any unwanted potential shift of the plate reference electrode(Fig. 2). As shown in Fig.3,although the sample solution was not stirred during the potentiometric measurement, the response of the ISEs was rapid. Measurements were made at 21 $\pm$ 0.5°C on a millivolt meter TOA Model HM-5BS or an on-line computer controlled ISE measuring system constructed in our laboratory(12). A typical experimental procedure is shown in Fig. 4.

RESULTS AND DISCUSSION

Na$^+$ and K$^+$ ions have certain degree of permeability of nonlysis origin through lipid bilayers. Therefore, those ions cannot be used for the marker ions, although Na$^+$ and K$^+$ ISEs were used along with liposomes to investigate the permeability of those ions through lipid bilayers(9,10). Among many candidate ions(11) which can potentially be trapped in liposome vesicles as marker ions, TPA$^+$ ion was finally chosen. The reason for this is that the molecular size of the TPA$^+$ ion is sufficiently large and is insoluble in lipid bilayer matrix so that the natural permeability of this ion through the liposome membranes is negligible during the course of the experiment (see Fig. 5). Also, since the selectivity coefficients of the TPA$^+$ ISE against Na$^+$ and K$^+$ ions are 5.0 x 10^{-6}, and 6.2 x 10^{-6}, respectively, the electrode interference from Na$^+$ and K$^+$ ions in serum was found negligible in contrast with the cases of K$^+$ and Na$^+$ ISEs.

Another difficulty in using the electrochemical method for studying immunological problems is that the amount of sample needed is generally large. However, this serious problem is also solved by making a very novel but simple experimental set up. We devised a "thin-layer potentiometric assembly"(3,4), where the microliter sample solution is held in a thin-layer space between the flat bottom of the ISE sensor and a plate-shaped silver/silver halide reference electrode(Fig. 1). With this setup, we could reduce the necessary sample volume three or four order of magnitude smaller than the ordinary ISE measurement without miniturizing the ISE itself. This assembly makes the electrochemical measurement of immunological system practically feasible.

The behavior of liposomes loaded with TPA$^+$ ion is found to be entirely consistent with the behavior of liposomes loaded with other markers(5-8). Fig. 6 shows the time course of the complement mediated release of TPA$^+$ ions by the cardiolipin antigen-antibody reaction on liposomes. The potentiometric response was found to be reproducibly dependent on concentrations of antiserum, antigen, and complement, and also on the reaction time for several different immunosystems(1,2). These results can be used for the immunoassay and also for the fundamental study of liposome immune lysis process. The ultimate sensitivity of this method for detecting antigens, antibodies, or the complement is high due to the amplification effect using liposome lysis process. At the present time, 10^{-11} M/L ganglioside GM_1 antigen was detectable in a volume of 50 μL.

The present method is the first successful electrochemical detection of liposome immune lysis process and is advantageous over the previous spectrophotometric, ESR, and fluorescence methods of liposome immunoassays in terms of by far simplified procedures, inexpensive fabrication, and easiness of continuous monitoring of the liposome immune lysis process, still having comparable sensitivity and precision. Another advantage of the present method comes from the use of the thin-layer micro-ISE, which made the necessary immuno sample volume as low as several microliters. This latter fact has completely overcome the barrier which the potentiometric method previously had for the measurement of precious immuno samples. Furthermore, it should be noted that the present method is also applicable to the study of liposome lysis process induced by such non-immunological origins as Sendai virus or C-reactive protein, which will be published soon from our laboratory.

REFERENCES

1. K.Shiba, T.Watanabe, Y.Umezawa, S.Fujiwara and H.Momoi, Chem.Letts., 1980, 155.
2. K.Shiba, Y.Umezawa, T.Watanabe, S.Ogawa and S.Fujiwara, Anal.Chem., 52, 1610 (1980)
3. K.Chiba, K.Tsunoda, Y.Umezawa, H.Haraguchi, S.Fujiwara and K.Fuwa, Anal.Chem., 52, 596 (1980)
4. Y.Umezawa and S.Fujiwara, Nippon Kagaku Zasshi, 1980, 1437.
5. J.A.Haxby, C.B.Kinsky and S.C.Kinsky, Pro.Nat.Aca.Sci.USA, 61, 300-307 (1968)
6. H.R.Six, W.W.Young, K.Vemura and S.C.Kinsky, Biochemistry, 13, 4050-4058 (1974)
7. Paul D.Orazio and G.A.Rechnitz, Anal.Chem., 49, 2083 (1977)
8. G.K.Humphries and H.M.McConell, Proc.Nat.Acad.Sci.USA, 71, 1691-1694 (1974)
9. C.W.M.Haest, J.DE Gier, G.A.Van Es, A.J.Verkley and L.L.M. Van Deenen, Biochem.Biophysica Acta, 288, 43-53 (1972)
10. A.Scarpa, J.De Gier, BBA, 241, 789-797 (1971)

11. Von R.Scholer and W.Simon, Helvetica Chimica Acta, <u>55</u>, 1801
    -1809 (1972)
12. K.Sawatari, Y.Imanishi, Y.Umezawa and S.Fujiwara, Bunseki
    Kagaku, <u>27</u>, 180 (1978)

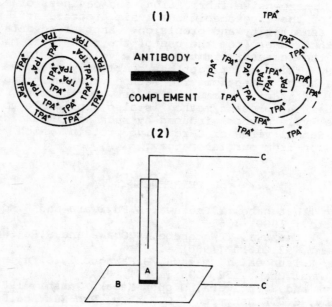

Figure 1. (1) Schematic drawing of complement mediated immune
         lysis process for tetrapentylammonium(TPA$^+$) ion
         loaded liposomes sensitized with appropriate lipid
         antigen, and resulting leakage of TPA$^+$ ions for
         analytical amplification.
         (2) Thin-layer potentiometric assembly for ISE
         measurement of microliter samples; A. TPA$^+$ ISE,
         B. Ag/AgCl plate reference electrode, and C. to a
         mV meter.

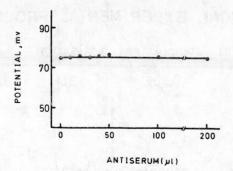

Figure 2. Influence of serum protein (rabbit) added on the observed potential of Ag/AgCl plate reference electrode vs. TPA^+ ISE; 5×10^{-5} M/L TPA^+ in Tris buffer saline; The volume of antiserum in the abscissa is already diluted one, and 50 and 100 microliters of antiserum, for example, correspond to 200 and 100 volume dilutions of stock antiserum, respectively.

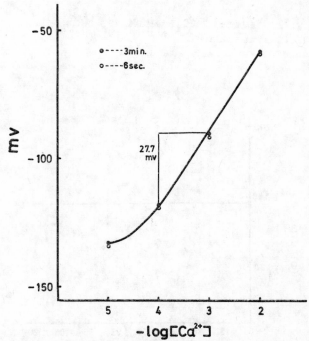

Figure 3. Dynamic(time-resolved) calibration curve for a Ca(II) ISE with the thin-layer potentiometric assembly ($10^{-2} - 10^{-5}$M $CaCl_2$ in0.1N KCl; A Ca(II) ISE was of Orion 93-20; 50 µL sample solution was used)

TYPICAL EXPERIMENTAL PROCEDURE

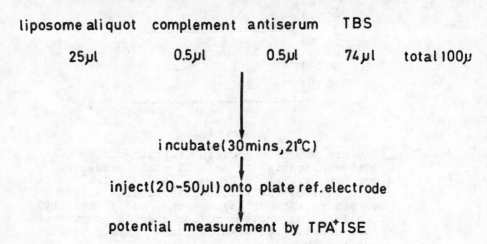

liposome aliquot   complement   antiserum   TBS

25μl   0.5μl   0.5μl   74μl   total 100μ

incubate(30mins,21°C)

inject(20-50μl)onto plate ref.electrode

potential measurement by TPA⁺ISE

Figure 4. A typical experimental procedure
          TBS: the modified Tris buffer saline(2).

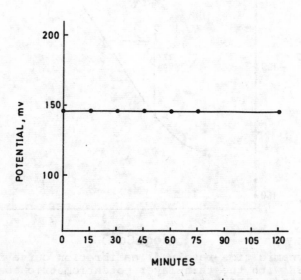

Figure 5. Continuous monitoring by a TPA$^+$ ISE of potential for
          the TPA$^+$ ion loaded liposome aliquat. The observed
          potential corresponds to ca. 5×10^{-5}M TPA$^+$.

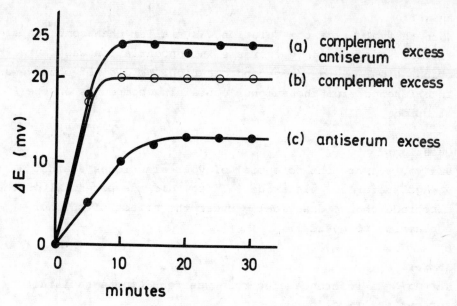

Figure 6. Time course of immune lysis process for cardiolipin-
lecithin sensitized liposome system; The antigen
and its rabbit antiserum were both supplied from
Iatron Co., Ltd.(Tokyo). Fresh complement(Guinea pig)
was a gift from Dr.T.Mukojima, National Cancer
Research Center(Tokyo).

QUESTIONS AND COMMENTS

Participants of the discussion: J:D.R.Thomas, E.Pungor,
L.Bartalits, G.Nagy, Y.Umezawa

Comment:
A comment on a question of terminology. It is not a micro-
-electrode that we have here but rather an ordinary electrode
dealing with micro-samples.

Question:
What is the effect of the volume of the injected sample on the
electrode response ? It is surprising that the sulphur-contain-
ing compounds do not interfere with the silver. What is your
comment on this ?

Answer:

When the volume of the solution was smaller than some ten micro litres, an appreciable increase in concentration was caused by evaporation. At to the second question: We did not observe any interference from the sulphur containing compounds on the electrode.

Question:

Did you measure the potential of your electrodes against a Ag/AgCl reference electrode ? If so, did the potential of the electrode change drastically under the effect of sulphur containing compounds ?

Answer:

The Ag/AgCl reference electrode was found to be satisfactory for our practical purposes.

Question:

How do you ensure that the serum-antiserum equilibrium point is attained during the measurements ?

Answer:

The optimum time of waiting was experimentally determined before the measurement. The optimum waiting time was 20 to 30 minutes.

Question:

The equilibrium point depends on the concentration of the serum. Do you dilute the serum before analysis ?

Answer:

Yes.

Question:

Is there any reason for your using TPA rather than any other tetraalkyl-ammonium system ?

Answer:

Tetramethyl and tetraethyl ammonium compounds were found to be unsatisfactory. TPA gave good results.

Question:

Do you see any way to increase the sensitivity of your method ?

Answer:

The potential step would be larger if we used liposomes. Unfortunately the latter is not sufficiently stable.

DYNAMIC CALIBRATION AND MEMORY EFFECT FOR VARIOUS KINDS OF ION-SELECTIVE ELECTRODES

Y. UMEZAWA, I. TASAKI AND S. FUJIWARA

Department of Chemistry, Faculty of Science, The University of Tokyo,
Hongo, Tokyo 113, Japan

ABSTRACT

Memory effect for ion-selective electrodes(ISEs) is the hysteresis of electrode responses caused by the sudden change of analyte concentrations to a lower direction. This memory effect was simulated systematically and quantitatively for the Cu(II) ISE using Cu(II) standard solutions. 1) The memory effect is increased with increasing the concentration difference between the samples of successive two runs. 2) When the concentration of the analyte is lower than 10^{-5} M, the memory effect is very serious even the concentration change is only of the order of one or two. 3) The memory effect influences the shapes of the calibration curves and makes them non-Nernstian, therefore,one has to be very careful to examine whether or not the memory effect is present for estimating the selectivity coefficient and the sensitivity of the electrode. 4) The response time is predominantly influenced by the presence of the memory effect, and it becomes extremely slow with the memory effect.

The memory effect for other ISEs were also examined and compared with that of Cu(II) ISEs: 5) Among several solid membrane ISEs examined, the F^- ISE exhibited the most serious memory effect, whereas the Cl^- ISE showed the least memory effect. 6) Among several liquid membrane type ISEs examined, K^+ ISEs showed almost no memory effect. On the contrary, the Ca(II) ISE exhibited the memory effect similar extent to the Cu(II) ISE, and 7) The memory effect for the gas sensing electrode such as an NH_3 electrode was very serious as expected.

These results will be useful for discussing the molecular feature of the memory effect and also for evaluating quantitatively the influence of the memory effect on the analytical measurement using ISEs.

INTRODUCTION

Memory effect for ion-selective electrodes(ISEs) is the hysteresis of electrode responses caused by the change of analyte concentration from high to a lower direction. This memory effect influences various aspects of ISEs, i.e., detection limit, response time, accuracy of analysis, estimation of selectivity coefficient, and applications to in situ and in vivo continuous analysis. We already examined the influence of the memory effect on the accuracy of the dynamic (time-resolved) calibration for the fluoride ISE using the NBS urine standard material for fluoride(1). Also, the memory effect for the Ca(II) ISE was systematically simulated using various combinations of Ca(II) standard solutions(2). The memory effect is expected to appear for a variety of ISEs including solid membrane, liquid membrane, and gas sensing electrodes, and, therefore, its quantitative evaluation is important.

In the present paper, the memory effect for various kinds of ISEs will be evaluated in terms of its influence on the dynamic calibration of the ISEs.

EXPERIMENTAL

In order to obtain precise and accurate time-resolved hard copies of the potential vs. time profiles of ISEs, a special experimental set-up was made(3). The analog output from the ISEs is first converted into frequencies followed by counting, serialization, and finally data aquisition by a PDP11/10 minicomputer. The precision for analog to digital conversion is 10 ppm.

ISEs used were of Denki Kagaku Keiki Co.(Japan), Orion (U.S.A.) and TOA Dempa Co.(Japan). Measurements were performed at $21 \pm 0.5°C$. Sample solutions made from deionized and distilled water were stirred by a magnetic stirrer. Chemicals used were of analytical reagent grade from Wako Co.(Tokyo).

Memory Effect for Cu(II) ISE

In the absence of the memory effect, namely, when the measurement is performed from low to higher concentrations, the response of the Cu(II) ISE is rapid, and the calibration curve constructed with time resolved responses at 6 s, for example, was already Nernstian to the same extent to that after 30 min (Fig.1). However, when the successive measurements are performed from a high to lower concentrations, the memory effect appears. Fig.2 shows the dynamic(time-resolved) calibration curves for a Cu(II) ISE in the presence of a 10^{-2} M memory: The measurement of each Cu(II) solution having the concentration range of 10^{-3} - 10^{-7} M was preceded by the measurement of a 10^{-2} M solution for 5 min., respectively. The electrode was rinsed and excess water was wiped off before and after each run. As shown in Fig.2, the memory effect is extremely serious at the concentratic range of 10^{-5} - 10^{-7} M. For a 10^{-7} M Cu(II) solution, apparent concentration measured at 6 s is still 5×10^{-5} M, indicating that the memory of 10^{-2} M solution remains at electrode surface as chemisorption, or physical adsorption and even after 6 s, these adsorbed ions are still partially on the electrode surface. Similarly, for the measurement of 10^{-6} M and 10^{-5} M solutions, the memory of the preceding measurement(10^{-2} M solution) remains at certain time durations. However, for the measurements of 10^{-3} M and 10^{-4} M solutions, the memory of a 10^{-2} M solution is very little and after 6 s, the calibration curve is already Nernstian. Fig.3 shows the dynamic calibration curves in the presence of the memory of 10^{-3} M solution. The results are very similar to those in Fig.2. Fig.4 shows the dynamic calibration curves with the memory of 10^{-4} M solution. The memory is appreciable for the region of 10^{-6} - 10^{-7} M. Fig.5 shows the dynamic calibration curve with 10^{-5} M memory. In this case, even the concentration change from the 10^{-5} M solution is of the order of one and two, respectively, the memory effect for 10^{-6} M and 10^{-7} M solutions is very enhanced. Fig.6 shows the time response of the electrode in 10^{-7} M solution with the memory of 10^{-6} M solution. In this case, the memory effect is serious even with the concentration change of the order of one.

Finally, the potential-time profile of the Cu(II) ISE with and without memory effect are shown in Fig.7.

Comparison with Other Electrodes

The memory effect for other types of electrodes including solid membranes, liquid membranes, and gas sensing electrodes are studied in the same way as those for Cu(II) ISEs.

Gas Electrodes

As an typical example for the gas sensing electrodes, an ammonia electrode was examined. As expected, the memory effect for an ammonia electrode is very serious and even after 30 min, the memory   still remains to a large extent(Fig.8). This result may be interpreted as follows: For the gas electrode, once the gaseous molecule comes inside the hydrophobic membrane, it almost irreversibly stays in for a long time duration. For discussing more quantitatively about the memory effect in the case of the gas electrode, various electrode designs such as the distance between the hydrophobic membrane and sensing glass electrodes underneath are being changed for further study.

Liquid Membrane ISEs

Several typical liquid membrane-type electrodes were examined. In the case of a Ca(II) ISE, the extent of the memory effect is overall very similar to that of the Cu(II) ISE(Fig.9). However, in the case of a potassium ion electrode, the memory effect is by far smaller than those of the Ca(II), and Cu(II) ISEs and is almost negligible(Fig.10). This dramatic difference between Ca(II) and K^+ ISEs is probably due to chemistry involved in the liquid ion exchange membrane rather than the differences of simple physical adsorption, and/or the structure of ISEs. For a NO_3^- electrode, the memory effect is not so serious as compared with the Ca(II) ISE, but still not negligible in contrast with the K^+ ISE.

Solid Membrane ISEs

Bromide ISEs exhibited slightly bigger memory effect than that of Cu(II) ISEs. Among several solid membrane type ISEs examined, the fluoride ISE exhibited the most serious memory effect in contrast with our initial expectation(Fig.11, and Table I). We first thought that the F^- ISE might show the least memory effect due to the fact that the membrane consists of LaF$_3$ single crystal which may have smooth surface with low porosity. It is interesting to note that the chloride ISE shows the least memory effect(Fig.12). This result again implies that the memory effect is not only a matter of simple physical adsorption or the like, instead, chemistry involved in the ion exchange process at the surface of the solid membrane also plays an important role for the mechanism of the memory effect.

Above results are all concerned with the standard solutions. It is expected that for actual analysis for river and sea water, and clinical samples, the memory effect may be much more enhanced due to the coexistence of some interfering ions and compounds.
It is noted that the extent of memory effect for commercial ISEs of the same kind was found to be very similar regardless of the manufacturers.

REFERENCES

1. Y.Umezawa, M.Nagata, K.Sawatari and S.Fujiwara, Bull.Chem. Soc.Jpn., 52, 241 (1979)
2. Y.Umezawa, I.Tasaki and S.Fujiwara, Nippon Kagaku Zasshi, 1980 , No.10 , p.1641
3. K.Sawatari, Y.Imanishi, Y.Umezawa and S.Fujiwara, Bunseki Kagaku, 27, 180 (1978)

Analyte pF / Memory pF	2	3	4	5	6	7
1	378 sec	426	912	840*	744*	1821*
2			195	1446*	3549*	3615*
3			333	2043	3549	4989
4				1917	2463	3306
5					1635	2358
6						1077

Table 1. Time needed for the recovery from memory effect for F$^-$ ISE. (Experimental condition same to those for Fig.11)

* Saturation of electrode response occurs at this time before the response reaches the final stage where the memory is completely removed.

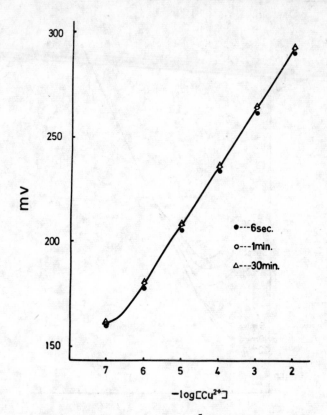

Figure 1. Dynamic /Time-resolved/ calibrations of a Cu/II/ ISE
in the absence of memory effect.
/10^{-2} - 10^{-7} M/L Cu/II/ ions in 0.1N KNO_3, the
volume of sample solution is 30 ml; The same also
true for Fig.2-Fig.7/.

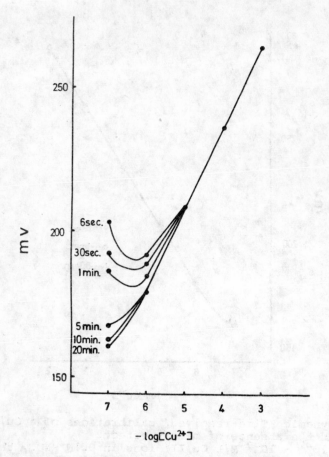

Figure 2. Dynamic calibrations of a Cu/II/ ISE in the presence of a "10^{-2} M memory" /see text:/.

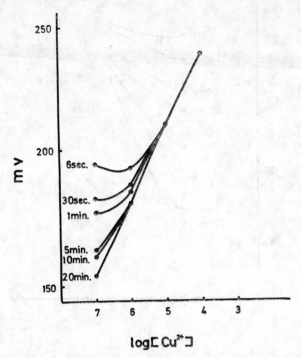

Figure 3. Dynamic calibrations of a CU/II/ ISE with a 10^{-3} M memory.

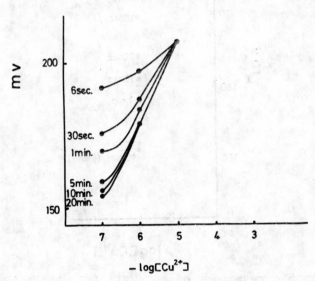

Figure 4. Dynamic calibrations of a Cu/II/ ISE with a 10^{-4} M memory.

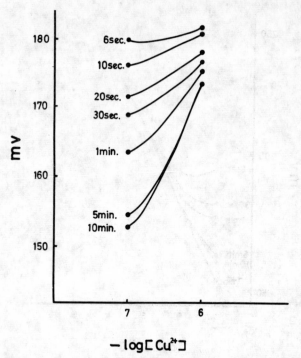

Figure 5. Dynamic calibrations of a Cu/II/ ISE with a 10^{-5} M memory.

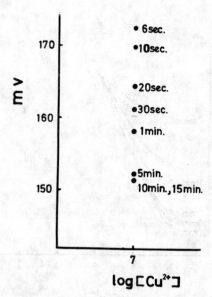

Figure 6. Dynamic responses of a Cu/II/ ISE with a 10^{-6} M memory.

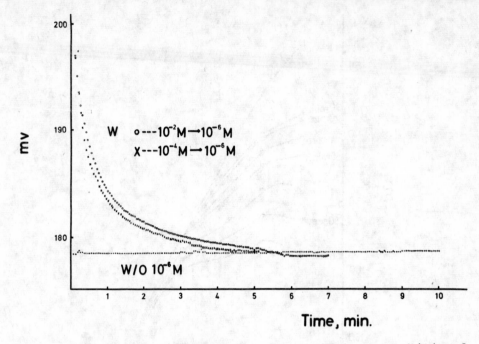

Figure 7. Potential-time profiles of a Cu/II/ ISE with/w/ and without/w/o/ memory effect.

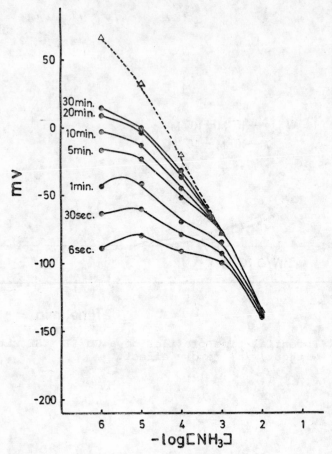

Figure 8. Dynamic calibrations of an NH_3 electrode in the presence of a $1 \times 10^{-1}M$ memory /10^{-1} - $10^{-6}N/L$ NH_4Cl in 0.1N NaOH/.

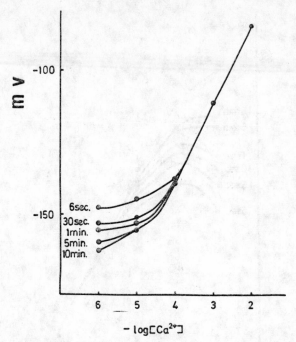

Figure 9. Dynamic calibrations of a Ca/II/ ISE with a 10^{-1}M memory /CaCl$_2$ in O.1N KCl/.

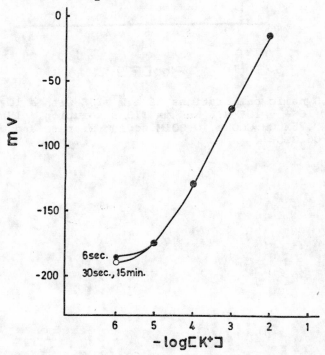

Figure 10. Dynamic calibrations of a K$^+$ ISE with a 1 x 10^{-1} M memory /10^{-1} - 10^{-6} M/L KCl in O.1N NaCl/.

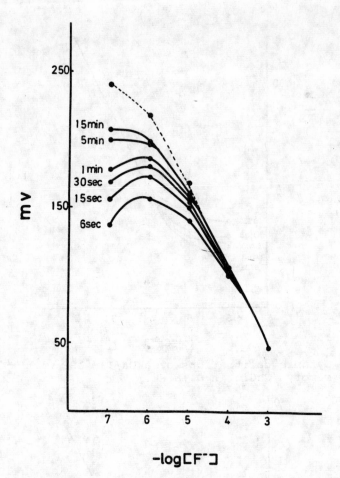

Figure 11.Dynamic calibrations of a F$^-$ ISE with a 10^{-1}M memory
/10^{-1} - 10^{-7}M/L NaF in TISAB solution / 1M NaCl,
0.75M CH$_3$COONa, 0.001M sodium citrate and 0.25M
CH$_3$COOH/.

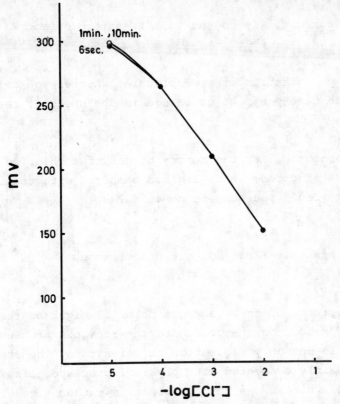

Figure 12. Dynamic calibrations of a Cl^- ISE with a $10^{-1}M$ memory /10^{-1} - 10^{-5} M/L KCl in 0.01N KNO_3/.

QUESTIONS AND COMMENTS

Participants of the discussion: W.Simon, L.Tomcsányi, A.Hulanicki, E.Pungor, G.Johansson, A.Lewenstam, Y.Umezawa

Question:
What you call memory effect is really a measure of the speed of response, the way you determined it. Could you comment on this ?

Answer:
Yes, your statement is correct.

Question:
If it is correct, why do you call it memory effect ?

Answer:
It is only a question of nomenclature, whether you call the phenomenon memory effect or change in the speed of response.

Question:
It was mentioned that the memory effect of a chloride ion-
-selective electrode is negligible compared with the bromide electrode. Could you comment on this ?

Answer:
No, unfortunately I can not give the reason.

Comment:
The composition of the electrode material might be responsible for this phenomenon. Returning to the previous problem, I think it is not only a question of definition. The response time is mainly connected with physical processes like the diffusion, whereas the memory effect is rather a result of a chemical change at the electrode surface or inside the membrane.

Answer:
The composition of our chloride and bromide electrodes was similar: 10% of silver chloride or silver bromide and 90% of silver sulphide.

Comment:
Some years ago we made response time measurements and explained the slower response by the so-called multi-electrode model, meaning that if about three quarters of the electrode surface were covered with the more dilute solution and one quarter with the more concentrated one, then the mixed potential meas-ured was very close to the electrode potential corresponding to the more concentrated solution. This means that if you make a large step-change in the solution activity in the direction of

dilution, a very small portion of the more concentrated solution remaining at the electrode surface can produce a memory effect, and this is a physical effect.

The other problem is that it is very difficult to reproduce the surface conditions, thus a relatively high number of experiments are needed to estimate the dynamic response characteristics of an ion-selective electrode.

Answer:
We have found the memory effect to be similar for electrodes manufactured by different companies. Electrodes from two Japanese companies and from Orion were studied.

Comment:
To come back to the speed of response and memory effect problem cases were shown where fast stirring almost or completely remo- ved the originally pronounced memory effect, indicating that it was really a response time problem. And there were several such cases shown during your lecture.

Comment:
If the electrode has some surface cracks, and a concentrated solution is trapped in the cracks, the exchange current density is higher at these places. Therefore the solution in the cracks will dominate the electrode potential. In this case the effect of the surface conditions is predominant, and the effect of stirring also seems to prove this.

Answer:
Yes, this is true.

Comment:
The memory effect in the case described in your paper appears to be mainly a physical, adsorption effect. However, physical and chemical processes are often mixed, for example when silver/I/ ions are reduced from silver iodide, iodide ions are adsorbed at the surface.

Question:

One can differentiate between memory effect /chemical pheno-
menon/ and response time /physical phenomenon/. The temperature
coefficients of the two effects should be different.

Answer:

I think this is a good idea. Temperature coefficients were
not measured so far, but I will try to do it later.

A FLOW INJECTION ANALYZER WITH MULTIPLE ISE-DETECTOR

RAUNO VIRTANEN

Technical Research Centre of Finland, Chemical Laboratory, Biologinkuja 7,
SF-02150 Espoo 15, Finland

ABSTRACT

A flow injection analyzer with a detector comprising up to
five flow-through ion selective electrodes was constructed.
Off-line automatic data acquisition was employed. The system
was used to analyses of mixed electrolyte solutions. The
concentrations of four ion species were measured simultaneously
with the four electrodes and the values obtained were corrected
with the aid of regression coefficients which were determined
by measurements of known mixtures.

INTRODUCTION

Ion selective electrodes (ISEs) are more or less sensitive
also to other ions than the primary one. In principle these
interferences could be accounted for and corrected by
mathematical methods but the selectivity coefficients usually
depend on experimental conditions and may vary rather much.
But if the composition of samples remains within certain
limits one can suppose that the coefficients remain constant.
Such a case occurs when a great number of one type of samples
are analyzed, e.g. serum analyses in clinical laboratories.
In such cases it can be thought that the selectivity
coefficients are measured for each electrode in known
solutions and the accuracy of analyses is improved by using
these coefficients in calculations /1/.

25*

Assume that electrode potential is measured in a reference solution (background solution in flow analysis) and in a standard solution, and the potential difference is denoted by ΔE, the slope of the calibration curve by S, and the concentration of the primary ion in the reference solution by c_r. For the mathematical treatment a quantity y is calculated /1/:

$$y = c_r \exp \frac{\Delta E}{S} \tag{1}$$

In a formal mathematical model the response function y of each electrode is described as a function of all other appropriate concentrations and also their combinations or cross interactions, e.g. for three ions:

$$y_1 = c_1 + k_2 c_2 + k_3 c_3 + k_{12} c_1 c_2 + k_{13} c_1 c_3 + k_{23} c_2 c_3 + {} \\ + k_{123} c_1 c_2 c_3 \tag{2}$$

Coefficient $k_2, \ldots, k_{123}$ can be determined by measuring the electrode potential in a series of solutions in which each ion occurres in two concentrations /2/. The concentration levels are chosen so that they cover the concentration range which occurres in samples. The higher and lower concentrations can be denoted by c_{+1} and c_{-1}, respectively. All the possible concentration combinations of different ions occur, so the number of test solutions is 2^N where N is the number of different ions. Owing to the calculation technique a coded concentration x, a pure number, is taken into use in equation (2):

$$x = \frac{c - \bar{c}}{(c_{+1} - c_{-1})/2} \tag{3}$$

where $\bar{c}$ is equal to $(c_{+1} + c_{-1})/2$. Equation (2) is now written as follows (for three interacting ions):

$$y = a_0 + a_1 x_1 + a_2 x_2 + a_3 x_3 + a_{12} x_1 x_2 + a_{13} x_1 x_3 +$$
$$+ a_{23} x_2 x_3 + a_{123} x_1 x_2 x_3 \qquad (4)$$

Constants a_i can be determined by measuring the electrode potential in the 2^N solutions (8 for three ions). Direct interactions of foreign ions to an electrode are expressed by the coefficients a_2 and a_3. The higher order coefficients reveal if there is any synergism between the ions, i.e. if the contributions of ions to their joint effect is different from their separate effects.

After the contants a_i have been determined the equation (4) can be used to calculate unknown concentrations. The potential of each electrode is measured in a sample. Thus an equation corresponding to equation (4) is obtained for each ion to be determined, e.g. four equations for four ions, from which the quantities $x_1, \ldots, x_4$ and further $c_1, \ldots, c_4$ can be determined.

EXPERIMENTAL

The multiple-electrode flow-through cell is shown in fig. 1. The electrode modules are machined from plastic so that they can be pressed together to form a serial cell through which the background solution can be pumped. The channel is 1 mm in diameter. The sample can be introduced into the cell in several ways. First, a four-way valve can be used. In fig. 2 there is shown a sample introduction system which resembles the injection block of a gaschromatograph. The sample is injected with a syringe through a rubber membrane into the flowing background solution. Fig. 3 shows an injection block which employs no membrane. Background solution flows into the injection block with a higher rate than drawn out by the pump through the cell so a part of the solution flows backwards out of the injection block. This model allows very easy injection of samples.

In this work four ISEs were used, potassium, sodium, calcium, and chloride selective electrodes. The sodium electrode was a commercial micro glass electrode (Micro-electrodes Inc. MI-420), the others were self-made. Potassium valinomycin and neutral calcium electrodes were of the PVC-membrane type /3/. The chloride electrode was made by melting a mixture of silver chloride and sulfide into a bar /4/ and drilling a hole through a 5 mm piece of the bar.

The data acquisition system was the same as in the earlier work /1/. It includes a five channel multiplexer with high impedance preamplifiers, AD-converter, control unit and an interface to a computer terminal. Data are collected off-line on magnetic tape, from which they can later be transferred to a computer for processing.

The solutions which were analyzed simulated the electrolyte composition of serum with respect to the four ions mentioned above. These ions do not cause mutual interferences in the concentrations at which they occur in serum but the correction method was applied to a case when the calcium and chloride electrodes were in rather poor condition in order to see if the results could be improved. All solutions contained 0.05 M triethylolamine and the pH was adjusted to 7.6 with nitric acid. The compositions of the solutions are given in table 1.

The flow rate of the background solution was 1 ml/min and the injected volumes of standards and samples were 50 µl.

RESULTS AND DISCUSSION

A recorder trace of the output of the potassium electrode during preliminary experiments is shown in fig. 4.

The measurements of the standard solutions gave values to the slopes of the electrodes and the coefficients in equation (4).

The slopes were: potassium electrode 57.6, sodium 56.4, calcium 13.3, and chloride -43.5 mV. Especially the calcium electrode was in poor condition. The regression equations were (only the significant coefficients are included):

$$y_k = 4.82 + 1.83 \ x_K$$
$$y_{Na} = 133 + 20.0 x_{Na}$$
$$y_{Ca} = 2.48 + 0.984 \ x_{Ca} -0.320 \ x_{Na} + 0.403 x_{Cl} \qquad (5)$$
$$y_{Cl} = 101.5 + 17.3 \ x_{Cl} - 1.36 \ x_{Na}$$

The results of the measurements of the sample solutions are given in table 2.

It is seen from equations (5) that the potassium and sodium electrodes needed no corrections as was expected and their results are quite satisfactory. On the other hand, the calcium electrode was effected by sodium and chloride ions and the chloride ions and the chloride electrode, in a less degree, by sodium ions. It may be pointed out that the correction coefficients are purely formal and they need not have any connection with the selectivity coefficients of the electrodes. The errors in the analyses of the samples were, however, reduced by using the regression equations, though errors in the calcium analyses still remained quite high.

So far the experimental material is too concise to allow definite conclusions to be made about the benefits of the correction method in this particular case.

REFERENCES

1. R. Virtanen, Conference on Ion-selective Electrodes, Budapest 1977, p. 589
2. O.L. Davies (Ed.), The Design and Analysis of Industrial Experiments, Longman Group Ltd., London, 2nd ed. 1978, p. 247

3. D. Ammann, R. Bissig, Z. Cimerman, U. Fiedler, M. Güggi,
   W.E. Morf, M. Oehme, H. Osswald, E. Pretsch, W. Simon,
   Proceedings of the International Workshop on Ion-Selective
   Electrodes and Enzyme Electrodes in Biology and Medicine,
   Urban and Schwarzenberg, München-Berlin-Wien 1976, p. 22

4. J. Siemroth, I. Hennig, R. Claus in E. Pungor (Ed.) Ion-
   Selective Electrodes, Akademiai Kiadó, Budapest 1977,
   p. 185

Table 1. Ionic concentrations of the solutions (mM)

	Background solution	Standard solutions lower higher level		Sample solutions			
		lower	higher	1	2	3	4
K^+	2.65	3.00	6.65	3.00	4.35	5.70	7.05
Na^+	100	113	153	155	110	140	125
Ca^{+2}	1.30	1.47	3.35	3.5	2.7	1.4	1.9
Cl^-	75	84	119	80	110	95	125

Table 2. Results of the analyses of the sample solutions

	K^+			Na^+			Ca^{+2}			Cl^-		
	a	b	c	a	b	c	a	b	c	a	b	c
1	3.00	1.3	-	155	-1.3	-	3.5	-40	-18	80	3.8	-2.5
2	4.35	-0.5	-	110	-4.5	-	2.7	20	-11	110	6.4	4.5
3	5.70	0.5	-	140	-1.4	-	1.4	-17.9	-6.4	95	2.1	2.1
4	7.05	-1.84	-	110	-0.8	-	1.9	34	13	125	6.4	5.6

a) Actual conc., mM
b) Relative error without correction
c)   "   " with correction

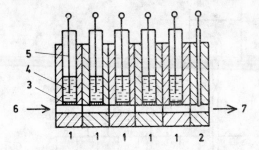

Fig. 1. The flow-through cell. 1. ISEs, 2. the reference
electrode 3. ISE membrane, 4. inner solution,
5. inner electrode, 6. solution in, 7. solution out

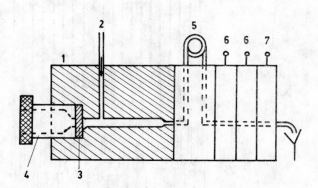

Fig. 2. Injection method 1. 1. Injection block, 2. background
solution from pump, 3. rubber membrane, 4. tightening
nut, 5. delay module, 6. ISEs, 7. reference electrode

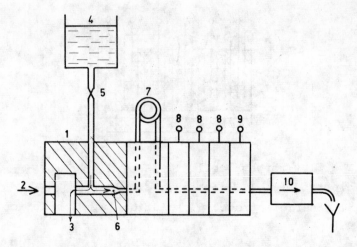

Fig. 3. Injection method 2. 1. Injection block, 2. sample in,
3. overflow, 4. background solution, 5. flow regulator,
6. injection point, 7. delay module, 8. ISEs,
9. reference electrode, 10. pump

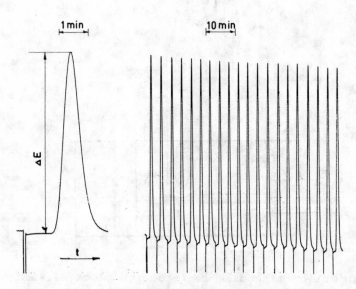

Fig. 4. Output of the potassium electrode. Same sample
injected repeatedly

QUESTIONS AND COMMENTS

Participants of the discussion: A.Hulanicki, J.D.R.Thomas,
E.Pungor, R.P.Buck, O.M.Petrukhin, R.Virtanen

Question:
Could you comment on the effect of chloride concentration
on the determination of calcium with calcium electrodes ?

Answer:
We did not study this phenomenon from a physico-chemical
point of view. The coefficients mentioned in our paper are
not selectivity coefficients but formal regression coefficients
that indicate all kinds of interferences.

Question:
Could you give some details of the type and prelife of your
calcium electrode ?

Answer:
A PVC neutral carrier-type electrode having a narrow linear
range of response to p_{Ca} has been used in our work.

Question:
According to your figure, 18 to 20 samples were measured in
an hour. Is this correct ?

Answer:
The results of preliminary tests with potassium electrodes
were shown in the figure. Later the rate of measurements was
slightly higher. The optimum rate of measurement will be
studied further in the future.

Question:
Could you comment on the fact whether or not it would be
advantageous to employ a small mixing chamber instead of the
long coil used in your experiment ?

Answer:
We have no experience as far as the use of a mixing chambre
is concerned.

Comment:
According to our experience it is advantageous to employ a
mixing chamber as the concentration profile is exactly defined
in this case. The interference of physical effects of the
flow in a capillary system can only be avoided by the use of
an integration type detection system such as the photometric
one.

Comment:
The equality of the calibration peaks is an evidence of the
reprodicibility of the measurements with the normal flow
injection method.

Answer:
A change in the viscosity or in the surface tension of the
flowing solution severely interfere with the measurement. The
calibration must be carried out under identical conditions
in such cases.

Comment:
I would agree with you when you are dealing with changes in
physical properties but I will not comment on the use of a
mixing chamber in this connection.

Comment:
Any type of system can be used in the analysis if the system
is physically well-defined, however, the mixing chamber is
preferable when changes in the physical parameters of the
system may occur.

Question:
How often do you calibrate your electrodes ?

Answer:

The system was not used for a period sufficiently long to permit an answer to this question. Our aim is to avoid frequent calibration.

Question:

Do you use a flowing electrolyte ? If so what is it ?
Do you use whole blood or plasma samples or do you consider to use them ? How accurate will be your measurements in these cases ?

Answer:

This work was carried out with inorganic synthetic solutions only. The system has not been tested in clinical analysis yet. We intend to carry out such measurements when the on-line data acquisition system will be in an operating condition. The electrolyte used in the experiments was an ethanolic buffer system.

INVESTIGATION OF THE ION SELECTIVITY MECHANISM
OF HYDROGEN ION-SENSITIVE FIELD EFFECT TRANSISTORS (ISFET)

YU. G. VLASOV, A. V. BRATOV AND V. P. LETAVIN

Department of Chemistry, Leningrad University, Leningrad, 199004, USSR

ABSTRACT

To obtain some information on the operation of pH-sensitive ISFETs (Ion-Sensitive Field Effect Transistors) high-frequency and quasistatic C-V measurements have been carried out on $Si-SiO_2$-electrolyte and $Si-SiO_2-Si_3N_4$ -electrolyte systems. It has been shown that both systems exhibited linear dependence of flat-band voltage with pH varying over the range from 2 to 10. The sensitivity of the oxide and the nitride structures is 30-40 mV/pH and 45-55 mV/pH, respectively.

The quasistatic C-V measurements showed that changes in C/C_{min} ratio of $Si-SiO_2$-electrolyte system, with pH varying could be attributed to the oxide charge fluctuations and/or to the changes in the surface states density at the Si/SiO_2 interface.

For $Si-SiO_2-Si_3N_4$ structures the correlation has been found between the pH-sensitivity of these structures and heat treatment in various ambients, which may lead to the decrease of hydrogen concentration in silicon nitride.

$Si-SiO_2-Si_3N_4$ structures have been found to be insensitive to 1 M excesses of Ca^{2+} and La^{3+} ions in solution and to the variation of the solution redox potential.

The ion sensitivity mechanism of such structures is very complicated, however, from the experimental results obtained it may be supposed that the major role is played by ion-exchange processes at the insulator/electrolyte interface.

INTRODUCTION

A great amount of papers devoted to the research and development of Ion-Sensitive Field Effect Transistors (ISFETs)./1,2/ have been published by now. As it is known ISFET is basically similar to a common field effect transistor except that an electrolyte and reference electrode immersed in it are used instead of metallic gate. In general conventional ion-selective (IS) membranes can be deposited on the top of gate insulator. In the case of pH-sensitive ISFET the role of IS membrane is played by the insulator itself either SiO_2 or Si_3N_4 /3 to 7/. It should be mentioned that devices with silicon nitride have better sensitivity and selectivity to hydrogen ions in solution.

387

It should be noted that although a lot of experimental data have been obtained concerning pH-sensitivity of ISFETs with silicon oxide and silicon nitride gate insulators there is no clear understanding of the exact ion-sensitivity mechanism. Some authors /7,8/proposed that the potential at the electrolyte/ insulator interface should be determined by the equilibrium

$$SiO^- + H^+ \rightleftharpoons SiOH \qquad (1)$$

which exists between H^+ ions in solution and the hydrated surface of the insulator (SiO_2 /8/, Si_3N_4 /7/).

For $Si-SiO_2$-electrolyte system Revesz /9/ and Bergveld /10/ assumed that hydrogen in some form is transported through the SiO_2 film and interacts with surface states at the Si/SiO_2 interface thus changing the surface states density.

Using ISFET one can obtain information about variations of the solution/ insulator interface potential when the composition of the solution changes. The same and sometimes more detailed information about the processes involved in electrolyte-insulator-semiconductor (EIS) systems can be obtained using capacitance - voltage (C-V) measurements /10,11/. According to Zemel /12/ a capacitor based on EIS system can also be regarded as Chemically Sensitive Semiconductor Device (CSSD).

This work has been carried out to obtain some additional information on the sensitivity mechanism of ISFETs with silicon dioxide and silicon nitride gate insulators.

EXPERIMENTAL

For $Si-SiO_2$ structures the starting materials were n-type ($\rho = 5\,\Omega\cdot cm$) and p-type ($\rho = 7.5\,\Omega\cdot cm$) silicon wafers (111) and (100) oriented respectively 500 - 2 500 Å thick layer of silicon dioxide has been obtained by standard thermal oxidation of the silicon substrate in dry oxygen at $1150^\circ C$.

For $Si-SiO_2-Si_3N_4$ structures n-type silicon wafers with main faces of (100) orientation and with resistivity of 4.5 or $7.5\,\Omega\cdot cm$ were used as substrates. 1000 Å thick layer of silicon dioxide was prepared in a manner described above. Chemically vapour deposited (CVD) silicon nitride films were obtained either by SiH_4-NH_3 or $SiCl_4-NH_3$ method in the presence of excess hydrogen. The temperature of deposition was $870-900^\circ C$. The structures prepared had different silicon nitride layer thickness (500-1200 Å).

ESCA analysis of Si_3N_4 films showed that some amount of oxygen was present at the surface and in the bulk of silicon nitride. The data obtained yield an approximate molar ratio of 3 silicon to 4 nitrogen.

The wafers were scribed into pieces approximately 4×4 mm^2 large. The back contact to silicon was made either by indium or by evaporated aluminium. The pieces were mounted in a sample holder according to /13/. The exposed to the solution area of insulator was in the oder of $1-2$ mm^2.

The quasistatic and high-frequency C-V measurements were carried out according to /14/ and /15/ respectively at the temperature of $25^\circ C$.

Fig.1 is representing a block-diagram of measuring set-up. The sample was placed into I M solution of KNO_3 or Na_2SO_4 and pH of the solution was changed by adding small amounts of appropriate acid or base. Solutions were prepared from analytical grade chemicals and bidistilled water. Small ac signal was applied between the sample and the platinum electrode with the total area of 1 cm^2. The dc bias was superimposed using the Ag/AgCl reference electrode (saturated KCl). The high-frequency measurements were carried out at I MHz.

According to /6/ the response and stability of an ISFET depends upon leakage current, I_L, which may flow through the insulator or across poorly encapsulated area. So we investigated devices which showed $I_L < 10$ nA under the superimposed bias.

THEORY OF OPERATION

EIS system has much in common with metal-insulator-semiconductor (MIS) structure. The capacitance of a MIS structure, C, is given by /16/:

$$\frac{I}{C} = \frac{I}{C_i} + \frac{I}{C_{sc} + C_{ss}}$$
(2)

where C_i is the geometrical capacitance of the insulator, C_{sc} –is the space charge capacitance and C_{ss} is the surface states capacitance of the semiconductor. They can be defined as

$$c_i = \frac{\varepsilon_o \varepsilon_i}{d} \quad ; \quad C_{ss} = \frac{\partial Q_{ss}}{\partial \mathcal{Y}_s} \quad ; \quad C_{sc} = \frac{\partial Q_{sc}}{\partial \mathcal{Y}_s} \quad ,$$
(3)

Here ε_o is the dielectric constant of vacuum, ε_i is the dielectric constant of the insulator, d is the thickness of the insulator, $\mathcal{Y}_s$ is the surface potential of the semiconductor, Q_{sc} and Q_{ss} are charges per unit volume in the semiconductor space charge region and in the surface states respectively. The effective voltage applied to the metal gate of the MIS structure, V_G, is given by the expression:

$$V_G = \mathcal{Y}_s + \mathcal{Y}_{ms} - \frac{Q_{ss}}{C_i} - \frac{Q_{sc}}{C_i}$$
(4)

where $\mathcal{Y}_{ms}$ is the metal-semiconductor work function difference. The flat - band conditions are provided for the semiconductor when

$$V_G = V_{FB} = \mathcal{Y}_{ms} - \frac{Q_{ss}}{C_i} .$$
(5)

The differential capacitance of an insulator-semiconductor structure in aqueous solution consists of the geometrical capacitance in series with the parallel combination of space charge capacitance, surface states capacitance and diffusion layer capacitance in the electrolyte /17/. We may choose the measuring conditions so that double layer and diffusion layer capacitance would be large enough to be neglected. Thus, like in MIS structures the total capacitance of an EIS system is determined by C_i, C_{ss}, and C_{sc}. When the metallic gate of the MIS structure is replaced by an electrolyte and reference electrode in it the expression for the gate voltage changes as follows:

$$V_G = \mathcal{Y}_s + E_{ref} + E + \mathcal{Y}_{ss} - \frac{Q_{ss} + Q_{sc}}{C_i} ,$$
(6)

where E_{ref} is the potential of the reference electrode, E is the potential difference between the bulk solution phase and insulator, $\mathcal{Y}_{ss}$ is the solution-semiconductor work function difference. For EIS system flat-band voltage is expressed by

$$V_{FB} = E_{ref} + E + \mathcal{Y}_{ss} \frac{Q_{ss}}{C_i} . \tag{7}$$

Thus we can see that if E_{ref}, $\mathcal{Y}_{ss}$ and Q_{ss}/C_i remain constant all changes in V_{FB} can be attributed to the changes in the potential difference E, which can be determined by any of the several processes of ion or electron exchange, adsorption, etc./13,18/. If charge exchange is assumed then E can be expressed by the Nernst equation:

$$E = E_o + \frac{RT}{zF} \ln a_i , \tag{8}$$

where E_o is a constant, R is a gas constant, T is the absolute temperature, F is the Faraday constant, z is the charge of species i, a_i is the activity of species i.

RESULTS AND DISCUSSION

Si - SiO$_2$ - electrolyte system

Experimental high-frequency C-V curves for Si-SiO$_2$-electrolyte system/19/ are presented in Fig.2. The parallel shift of these curves along the voltage axis is due to the pH changes of electrolyte. With the help of nomogramms /II/ one can calculate the "ideal" C-V curves thus determining the flat - band capacitance, C_{FB}, of the structure (see Fig.2). Knowing C_{FB} we can obtain the value of flat-band voltage, V_{FB}, for the EIS system with the definite pH value from the experimental C-V curves. Fig.3 shows V_{FB} of the Si- -SiO$_2$-electrolyte system as a function of pH. It can be seen that this function is linear over the pH range from 2 to 10. The sensitivity of these devices is approximately 30 mV/pH. In solutions with pH 0.9 - 2.0 and 10 - 12 irreversible changes in V_{FB} occured accompanied by the increase of the geometrical capacitance of silicon dioxide which was more pronounced in basic solutions.

Fig.4 presents a family of quasistatic C-V curves of Si-SiO$_2$-electrolyte system at different pH values. The shift of these curves is due to the change of the electric potential difference across the SiO$_2$/electrolyte interface produced by changes in pH. It also can be seen that different pH values of the electrolyte lead to differences in C/C_{min} ratio. According to/9, 10/ this can be attributed to the changes of the surface states charge at Si/SiO$_2$ interface. Calculations made according to Berglund formulas /6/ give $N_{ss} = 1.5 \times 10^{10}$ cm^{-3} for pH = 1.5; $N_{ss} = 2.0 \times 10^{10}$ cm^{-3} for pH = 6.0; and $N_{ss} = 2.5 \times 10^{10}$ cm^{-3} for pH = 11.0. On the other hand, the analysis of experimental C-V curves by the method suggested in /20/ shows that the same effect may also result from the oxide charge fluctuations.

If we assume that pH variations of the electrolyte induce changes in surface states charge then two terms in equation (7), E and Q_{ss}, will be pH-dependent. The data obtained show that the response time of Si-SiO$_2$-electrolyte system to the change of pH is less then one second and is independent of the SiO$_2$ layer thickness. So it is unlikely that primary potential generating process should involve the transport of some H-bearing species through the SiO$_2$ film as it was suggested by Revesz /9/ and Bergveld /10/. Undoubtadly diffusion may occure in this type of systems but it

reqires much more time and it would result in long term drift of the flat -
band potential. Such slow drift, for example, had been found by Leistico /21/.

As it was noted by many investigators /1-3/ Si-SiO$_2$ structures have low
sensitivity compared with the theoretical Nernstian response. It's no wonder,
for many silica glasses show similar behaviour /22/.

Si-SiO$_2$-Si$_3$N$_4$-electrolyte system

As it was mentioned above silicon nitride had been obtained by two different methods. Si$_3$N$_4$ produced by ammonolysis of monosilane had been subjected
to the heat treatment in vacuum, argon, and hydrogen ambients. Now it is obvious that a great amount of hydrogen is present in CVD films of silicon nitride /23,24/ in the form of NH and SiH groups and that the concentration of
these groups can be regulated by varying deposition and annealing temperatures and NH$_3$/SiH$_4$ ratio. Concentration of SiH and NH groups was estimated
with the help of multiple internal reflection (MIR) technique. Initial Si$_3$N$_4$
layer contained 2 10^{22} cm^{-3} NH groups. The concentration of SiH groups was
too small to be detected. Annealing in argon and vacuum (10^{-2} Torr) reduced
the concentration of NH groups to 3 10^{21} cm^{-3}. Annealing in hydrogen did not
change initial concentration of these groups.

Fig.5 shows the flat-band voltage of the Si-SiO$_2$-Si$_3$N$_4$-electrolyte system
with initial Si$_3$N$_4$ as a function of pH of the electrolyte. It can be seen
that a linear dependence of V_{FB} vs pH exist over the pH range from 2 to 10.
The sensitivity of these devices is 50±5 mV/pH. Repetitive measurements
showed that devices exhibited some unstability of V_{FB}. After annealing in
hydrogen ambient during 1 hour at 1000°C devices became more stable and
showed good reproducibility of the results, while annealing at 1000°C during
1 hour in vacuum or argon ambients made all characteristics of the devices
worse: sensitivity decreased to 25-30 mV/pH and the hysteresis developed in
the V_{FB} vs pH dependance as shown in Fig.6. The nature of this hysteresis
is not yet clear.

Etching of the samples in diluted (1:100) solution of HF acid in order
to remove thin layer of SiO$_2$ which might appear during heat treatment or
storage did not cause any changes in device sensitivity.

Devices with silicon nitride obtained by ammonolysis of tetrachlorosilane
exhibited stable pH response with the sensitivity of 50±5 mV/pH and the
range of linear V_{FB} vs pH dependance from pH 2 to pH 10.

From the electrochemical point of view electrolyte/solid interface may
be described as blocked (or polarized) or unblocked (or nonpolarized)/25/,
however some intermediate case is more realistic. If we assume that Si$_3$N$_4$/
/electrolyte interface is totaly blocked (all processes of charge exchange
are absent) then term E in equation (7) represents the potential drop in
double layer at the electrolyte/solid interface. So in that case changes in
double layer parameters will induce changes in V_{FB} of the EIS system.

We performed special tests in which Ca^{2+} and La^{3+} ions were added to the
solution. Presence of these ions in solution should change the charge distribution in double layer drastically but up to 1 M concentrations of these
ions in solution the flat-band voltage of the system was unchanged.

According to Brouwer /26/ changes in the redox potential of the solution may lead to the changes of Fermi level in the solution and thus alter
the work function difference between semiconductor and solution. Measurements carried out in solutions containing K$_3$Fe(CN)$_6$ and K$_4$Fe(CN)$_6$ in ratios
1:10, 1:1, and 10:1 at pH = 2.4 show that Si-SiO$_2$-Si$_3$N$_4$-electrolyte system

is insensitive to the changes of the redox potential of the solution.

To sum up, the purpose of this paper was to show that investigation of the ion sensitivity of ISFETs, which require for their development rather expensive and specific equipment, can be carried out using more simple capacitor like electrolyte-insulator-semiconductor system.

The data presented show that $Si-SiO_2-Si_3N_4$ structure behaves in the simple Nernstian manner in response to the hydrogen ion activity. Heat treatment in different gas ambients and vacuum, which may result in dissociation of NH groups in Si_3N_4 and some structural changes in it, has pronounced influence on the pH response of these devices. The silicon nitride system was shown to be insensitive to the presence of multivalent ions in solution and to the changes of the redox potential of the solution.

REFERENCES

1. Yu.G.Vlasov, Zh.Prikl.Khim. 52, 3 (1979)
2. J.Janata, Ion-Selective Electrode Rev. 1, 31 (1979)
3. P.W.Cheung, W.H.Ko, D.J.Fung, and S.H.Wong, in Theory, Design, and Biomedical Application of Solid State Chemical Sensors (P.W.Cheung et al.ed.) CRC, Cleveland, 1978
4. P.Bergveld, IEEE Trans Bio-Med.Eng. BME-17, 70 (1970)
5. P.Bergveld, IEEE Trans Bio-Med.Eng. BME-19, 342 (1972)
6. S.D.Moss, J.B.Smith, P.A.Comte, C.C.Johnson, and L.Astle, J.Bioeng.1, 11 (1977)
7. M.Esashi and T.Matsuo, IEEE Trans. Bio-Med.Eng. BME-25, 184 (1978)
8. J.F.Schenk, J.Colloid Interface Sci. 61, 569 (1977)
9. A.G.Revesz, Thin Solid Films 21, L43 (1977)
10. P.Bergveld, in Physics of SiO_2 and its Interfaces (S.T.Pantelides ed.) Pergamon Press. New York 1978
11. K.H.Zaininger and F.P.Heiman, Solid State Technol. May, 49; June,46 (1970)
12. J.N.Zemel, Anal.Chem. 47, 255A (1975)
13. R.P.Buck and D.E.Hackleman, Anal.Chem. 49, 2315 (1977)
14. M.Kuhn, Solid State Electronics 13, 873 (1970)
15. K.H.Zaininger, RCA Rev. 27, 341 (1966)
16. S.M.Sze, Physics of Semiconductor Devices, Wiley, New York, 1969
17. A.Wolkenberg, Phys.Stat.Sol.(a) 48,203 (1978)
18. J.Janata, in ref.3.
19. Yu.G.Vlasov,Yu.A.Tarantov, A.P.Baraban, and V.P.Letavin, Zh.Prikl.Khim. 53,1980 (1980)
20. V.G.Litovchenko and A.N.Gorban, Osnovi Fisiki Mikroelectronnikh Sistem metall-dielectric-poluprovodnik, Naukova Dumka, Kiev, 1978.
21. O.Leistico, Physica Scripta 18, 445 (1978)
22. N.Lakshminarayanaiah, Membrane Electrodes, Academic Press, New York,1976
23. H.J.Stain, S.T.Picraux, and P.H.Hollway, IEEE Trans. Electr.Dev. ED-25, 1008 (1978)
24. V.I.Belyi, F.A.Kuznetsov, T.P.Smirnova, L.V.Chramova, and L.Kh.Kravchenko, Thin Solid Films 37, L39 (1976)
25. J.O`M.Bockris and A.K.N.Reddy, Modern Electrochemistry, A Plenum/Rosetta Edition, New York, 1970
26. G.Brouwer, J.Electrochem.Soc. 114, 743 (1967)

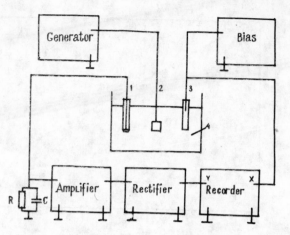

Figure 1. Block-diagram of the experimental set-up for high-frequency
C-V measurements. 1) the sample, 2) the platinum electrode,
3) the reference Ag/AgCl (saturated KCl) electrode

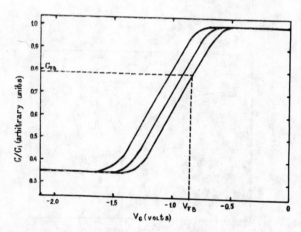

Figure 2. Experimental high-frequency C-V curves of the Si-SiO$_2$-elec-
trolyte system at different pH values of the electrolyte.
C_{FB} and V_{FB} are the flat-band capacitance and flat-band voltage respectively

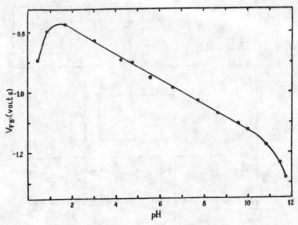

Figure 3. Flat-band voltage of the $Si-SiO_2$-electrolyte system as a
function of pH

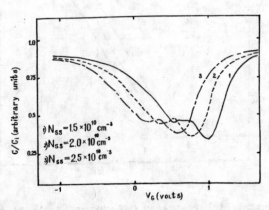

Figure 4. The quasistatic C-V curves of the $Si-SiO_2$-electrolyte system
at different pH values of the electrolyte. 1) pH = I.5,
2) pH = 6.0 , 3) pH = 11.0 . N_{ss} is the surface states density
calculated in C/C_{min} point

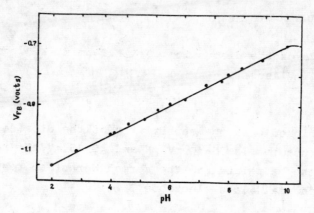

Figure 5. pH response of the Si-SiO$_2$-Si$_3$N$_4$-electrolyte system with initial silicon nitride ($C_{NH} = 2 \ 10^{22} \ cm^{-3}$)

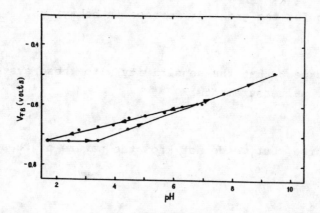

Figure 6. Hysteresis in the V_{FB} vs pH dependence . (Device with Si$_3$N$_4$ after heat treatment at 1000 ^{o}C during 1 hour in argon ambient $C_{NH} = 3 \ 10^{21} \ cm^{-3}$)

QUESTIONS AND COMMENTS

Participants of the discussion: J.D.R.Thomas, K.Burger,
T.Fjeldly, Y.Vlasov

Question:
It was mentioned that the sensitivity of the dioxide and
nitride structures is 30-40 mV/pH and 40-45 mV/pH, respectively.
How can these be related to the normal Nernstian response we
get from a normal glass electrode ?

Answer:
In this case a slope lower than Nernstian was observed. This
may be due to some processes taking place in the oxide or
nitride of silicon.
   Many research groups are working on the project to find
out which kinds of processes are responsible for the electrode
function of ISFET-s, where a silicon oxide layer is formed.
However, I do not believe that these two systems can be
compared. In the case of ISFET-s there is a very thin dielectric film. The properties of this film are quite different
from those of the classical electrode materials. Some special
processes like diffusion, changes in the charge state may be
responsible for the great difference in behaviour.

Question:
Would you expect that the sensitivity with these systems can
be used for pH measurements ?

Answer:
Yes, of course, but we do not know the nature of these
differences.

Question:
By listening to the lecture, I had the idea that the anion
effect may be due to the formation of different species in
the surface layer. This very small surface layer differs from
the bulk in dielectric properties. You have nitrate and

chloride, and different species could be formed. Do you think that this fact can be responsible for the anomalous behaviour ?

Answer:
The information we extracted from CD measurements led us to the conclusion that there are changes at the interface of the dielectric film and the solution. It is also possible that there are some changes in the surface charge.
Detailed investigations of these processes are ahead.

Question:
From the book by Matsuo and his coworkers on the oxide and nitride layers on ISFET-s it seems that silicon dioxide is not suitable at all for pH measurements, due probably to hydrolytic processes. The nitride and aluminium oxide are very good, but the nitride works properly only if it is free of oxygen.

Answer:
Matsuo is a pioneer in this field. According to him the best material is silicon nitride. Silicon dioxide is not selective since it responds to protons as well as to sodium, potassium and other ions. Silicon dioxide is interesting only from the theoretical point of view.

Silicon nitride makes the best dielectric film. It is sensitive to protons and has a high selecetivity. The problem still exists: what is the reason for the pH-sensitivity of silicon nitride. It was found that the actual silicon nitride film has some N-H and Si-H bonds. The real bonding in this film is not Si/III/ and Si/IV/. You get the nitride film on the surface in nitrogen atmosphere at high temperature, and the surface concentration of these N-H bonds is quite large. You can never get silicon nitride by etching, you have always about 3-5% of silicon oxide in it, as reported by several authors.

INVESTIGATIONS OF LIQUID MEMBRANES BASED ON CHELATES OF COPPER(II) WITH DIALKYLDITHIOPHOSPHORIC ACIDS IN DIFFERENT ORGANIC SOLVENTS

V. A. ZARINSKII, L. K. SPIGUN, G. E. VLASOVA, I. V. VOLOBUEVA, E. V. RYBAKOVA
AND YU. A. ZOLOTOV

V. I. Vernadsky Institute of Geochemistry and Analytical Chemistry, USSR Academy of Sciences,
Moscow, USSR

ABSTRACT

The behaviour of chelates of copper/II/ with dialkyldithio-phosphoric acids type $[C_nH_{2n+1}O/_2P/S/S]_2Cu$ /n=20-24/ as an active material in liquid-membrane electrode is described. Nernstian response is obtained over the pCu range of 1.0-5.0 with the slope of /29.2±0.5/ mV/pCu at an ionic strength of 0.1 and 22°C in pH range from 3.5 to 6.0. Selectivities for copper/II/, with respect to silver/I/, mercury/II/, lead/II/, zinc/II/, cadmium/II/ and other interfering species as well as pH response have been examined.

The practical response times of the membrane electrode have been studied. The influence of the solvents on the response characteristics of the liquid membranes has been also investigated. Some possibilities of application of the newly described liquid-membrane electrode, sensitive to Cu/II/-ions for direct potentiometric measurements are also presented.

INTRODUCTION

The application of metal chelates as active components in liquid-membrane electrodes offers the exciting prospect for the preparation of ion-selective electrodes with the optimal parameters.

It has been pointed out that the potentiometric selectivity

of a liquid-membrane electrode is considerably dependent on solvent extraction constants /1/. On this basis dialkyldithiophosphoric acids which have the high extractive selectivity to bivalent transition elements were found to be suitable for using in liquid- and PVC-matrix-membrane electrodes /2,3/.

In the present lecture the results of the potentiometric investigations of some liquid Cu-sensig membranes based on the chelates of copper/II/ with dialkyldithiophosphoric acids type $[/C_nH_{2n+1}O/_2P/S/S]_2Cu$ /n=20-24/ in different organic solvents are described.

EXPERIMENTAL

E.M.F. measurements have been performed with a digital voltmeter type Ц 1513 /USSR/, which has been coupled with a priater type Ф 5033K /USSR/.

The investigated cell had the following scheme:

| Ag/AgCl | Internal solution | Liquid membrane | Sample solution | Electrolyte bridge | KCl/satd./, AgCl/Ag/I/. |

Liquid membranes consisted of a 8.10^{-3} M solution of dialkyl-dithiophosphate complexes of copper/II/ in different organic solvents /chloroform, n-decanol, 1,2-dichloroethane, chloro-benzol, chlorex/.

A caprolon body has been used in the electrode of own construction. The general principle was similar to that of the Orion liquid-membrane electrode, i.e. an internal aqueous solution separated from an aqueous sample solution by means of a thin porous matrix saturated with an organic solution. The internal filling solution was aqueous 0.01 M $CuSO_4$ and 0,01 M KCl in a saturated solution of AgCl.

The membrane electrode potential was measured against a silver/silver chloride electrode, incorporating a 0.1 M Na_2SO_4 electrolyte bridge, type 90-02 /Orion, USA/.

The sample solutions were stirred magnetically and thermostated at 22 ± 1^{O}C.

RESULTS AND DISCUSSION

The dependence of the e.m.f. of the cell /1/ on the copper /II/ concentration was investigated in the copper sulphate solutions in the range from 10^{-1} to 10^{-6} M containing 0.1 M sodium sulphate, 0.1 M sodium nitrate, 0.1 M and 2.0 M potassium chloride or 0.05 M acetate buffer /pH 4.5/. Figure 1 shows that in the presence of 0.1 M sodium sulphate, sodium nitrate and potassium chloride the liquid membrane provides approximately Nernstian response over the pCu range of 1-5. The slope of the linear range of the calibration curve is /29.2+0.5/ mV/pCu at 22°C. As can be seen in Figure 1 there are no large differences between the behaviour of the liquid membrane in these solutions and thus the nitrate, sulphate and chloride anions do not affect the e.m.f. of the cell /1/. The limit of detection determined according to the definition by IUPAC /4/ is pCu 5.3 . However at the total chloride concentration higher than 1 M the limit of detection increases to pCu 3.9 and the slope is only about /25.5+0.7/ mV/pCu. In the medium of acetate buffer /0.05 M, pH 4.5/ the membrane shows the linear response to copper/II/-ions in pCu range of 1.0-3.3 with the slope of /23.0$\pm$0.5/ mV/pCu. These results are caused by the decrease of true activity of copper/II/-ions due to the interaction between the copper/II/-ions and acetate or chloride anions in the aqueous solution.

The influence of pH over the range of 2-8 on the e.m.f. of cell /1/ has been tested for solutions containing 10^{-2}-10^{-4}M copper sulphate. The results obtained are given in Figure 2. The membrane electrode potential is constant in the pH range of 3.5-6.0. It is limited by the hydrogen ion concentration in the acidic solution and at the higher pH region by the formation of hydroxo complexes and precipitation of copper hydroxide.

The practical response time of the liquid membrane electrode depends on the copper/II/ concentration and composition of the sample solution. The results /Figure 3/ show that it varies from 30 s in concentrated pure copper/II/ solutions /10^{-1} M/ to about 1 min in diluted solutions /10^{-4} M/, but at a concentration less than 10^{-5} M the practical response time increases to 5 min.

The potentiometric selectivity coefficient values of liquid-
-membrane electrode calculated by the separate solution method
/4/, are presented in Table 1. These values show that the
membrane has a high selectivity for copper/II/ compared to
the alkali and alkaline earth metals. At constant ionic strength,
variations in the concentration of these cations do not affect
the membrane electrode potential. The sequence of the potentiometric selectivity for copper/II/ in respect to other bivalent
transition metals is following: Cu/II/ > Pb/II/ > Cd/II/ > Zn/II/ >
Mn/II/ > Co/II/ ≃ Ni/II/ and generally agrees with that of the
stability constants of the corresponding metal-dialkyldithio-
phosphate chelates /5/. The cations Pd/II/, Hg/II/ and Ag/I/
form strong complexes with the dialkyldithiophosphoric acids
/6/ and so they have the strongest effect on the e.m.f. of cell
/I/.

The experimental results indicated an important role of
membrane solvents in the response characteristics of the
liquid membrane. It has been stated that the changes in solvents of liquid membrane can lead not only to kinetic exchanges
but also to the ion transport mechanism through the membrane
and the membrane - sample solution interface. The strong effect
of the different membrane solvents on the e.m.f. of cell /I/
is illustrated in Figure 4. In all cases the e.m.f. is linearly
dependent on the logarithm of the activity /concentration/ of
copper/II/-ions with a nearly theoretical slope of 29 mV/pCu
at 22°C, but the detection limit increases in the following
order: 1,2-dichloroethane < n-decanol < chlorobenzol < chloroform < chlorex. This is due to the fact that the solvents
affect the solubility parameters of the membrane components.
Some solvents may have also their own response to various
anionic or cationic species.

The membrane solvent has also an effect on the response
time of the membrane electrode by influencing the rate of
the ion-exchange reaction between the electrode membrane and
the sample solution as well as the transfer numbers of the
ionic species in the membrane phase. The response time may
be also limited by a slow process of dissolution of the mem-

brane components including organic solvents, in the aqueous sample solution.

From Figure 5 it can be seen that the liquid membranes containing chlorex and chlorobenzol as solvents have about the same response times. In contrast, 1,2-dichloroethane membrane electrode shows markedly /about 5-times/ slower response than other membranes.

The solvent may affect membrane potentiometric selectivity to copper/II/ ions due to the changes in the formation constants and partition coefficients which determine ion--exchanger equilibria between the copper/II/ ions in the liquid membrane and the other cations in the sample solution. The solvent integration changes in selectivity are well balanced. This is illustrated in Figure 6. The potentiometric selectivity coefficients for Cu/II/ ions in respect to Pb/II/ ions $pK_{Cu/Pb}$ seems to be rather independent of the dielectric constants of the membrane solvents. In contrast, the $pK_{Cu/Mn}$ and $pK_{Cu/Co}$ values gradually increase with the decreasing of the dielectric constant.

In summary we may conclude that a new liquidstate copper/II/ -selective electrode has been developed. It has been used with good efficiency for potentiometric determination of copper/II/ concentration both directly in acetate buffer and by the standard addition method. Some examples are presented in Table 2.

REFERENCES

1. S. Back, J. Sandblom, Anal.Chem., 45, 1680 /1973/
2. W. Szczepaniak, K. Ren, Anal.Chim.Acta, 82, 37 /1976/
3. E.A. Materova, V.V. Mychovikov, M.G. Gregoreva, Ion-exchange and Ionometry v.2, Leningrad, USSR, 1979, p. 142
4. G.G.Guilbault, Ion-Selective Electrode, Rev. 1, 139 /1979/
5. D.H. Handly, I.A.Deap, Anal.Chem., 34, 1312 /1962/
6. V.F. Toropova, Obshch.Khim.,40, 1043 /1970/; 41, 1673 /1971/

Table 1. Potentiometric selectivity coefficients obtained
by the separate solution technique using 0.01 M
solutions at ionic strength 0.1; t = 22°C

Membrane solvent : 1,2-dichloroethane

Interfering cation	$pK_{Cu/Me}$ n=5; p=0.95	Interfering cation	$pK_{Cu/Me}$ n=5; p=0.95
Ag/I/	-0.14±0.04	Co/II/	2.65±0.14
Hg/II/	-0.34±0.06	Mn/II/	2.00±0.07
Pb/II/	0.53±0.06	Cd/II/	1.14±0.09
Zn/II/	1.37±0.08	Ca/II/	3.78±0.27
Ni/II/	2.71±0.12	Na/I/, K/I/	4.20±0.44
Pd/II/	-0.87±0.08		

Table 2. Potentiometric determination of Cu/II/ in
different solutions

Sample	C_{Cu}, mg/ml taken	found	Error,%	Stand. dev.
	0.020	0.019	5.0	0.002
	0.060	0.063	5.0	0.005
CuSO$_4$-	0.100	0.097	3.0	0.006
NaCl/0.5M/	0.508	0.520	2.4	0.020
	2.540	2.590	2.0	0.091
	5.080	4.960	2.3	0.110
CuSO$_4$-	0.050	0.052	4.0	0.003
NiSO$_4$/0.01M/	0.100	0.096	4.0	0.006
CaCl$_2$/0.01M/	0.510	0.502	1.6	0.012
CuSO$_4$-	0.100	0.095	5.0	0.008
ZnSO$_4$/0.01M/	0.510	0.528	3.5	0.036
	2.560	2.621	2.4	0.094
CuSO$_4$-	0.050	0.053	6.0	0.003
CaCl$_2$/0.1M/	0.100	0.096	4.0	0.007
	0.570	0.500	2.0	0.015
CuSO$_4$-	0.100	0.106	6.0	0.009
CdCl$_2$/0.01M/	0.510	0.530	3.9	0.028
	5.080	5.210	2.6	0.171

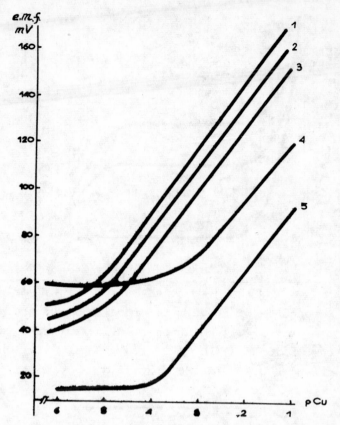

Fig. 1   E.m.f. response of cell/I/ to aqueous solutions
$CuSO_4$ in the presence of 0.1 M Na_2SO_4 /1/; 0.1 M $NaNO_3$
/2/; 0.1 M KCl /3/; 0.05 M acetate buffer, pH=4.5 /4/
and 2 M KCl /5/

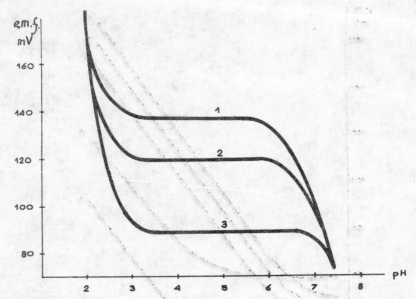

Fig. 2   Dependence of the e.m.f. of cell /I/ on the pH of
the pure sample solution at different $CuSO_4$ concentrations. 1- 10^{-2}; 2- 10^{-3}, 3- 10^{-4} M. Ionic
strength 0.1 M /Na_2SO_4/; t=22°C

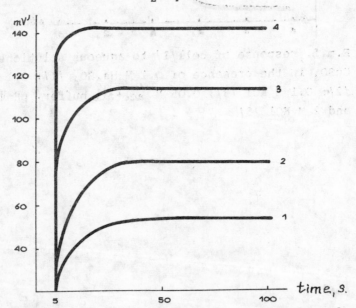

Fig. 3   Dynamic response of the Cu-selective liquid membrane
in pure $CuSO_4$ solutions. Concentration of $CuSO_4$:
1- 10^{-4}; 2- 10^{-3}; 3- 10^{-2}; 4- 10^{-1} M

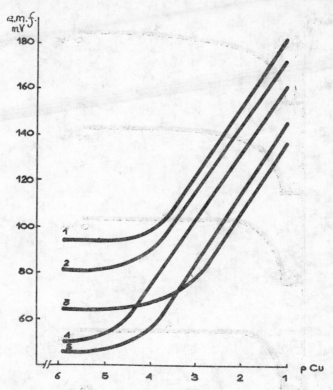

Fig. 4   E.m.f. response of cell/I/ versus pCu for the liquid
membranes based on $\left[/C_nH_{2n+1}O/_2P/S/S\right]_2$ Cu /n=20-24/
in different organic solvents.
1- chloroform; 2- chlorobenzol; 3- chlorex; 4- 1,2
dichlorethan; 5- n-decanol

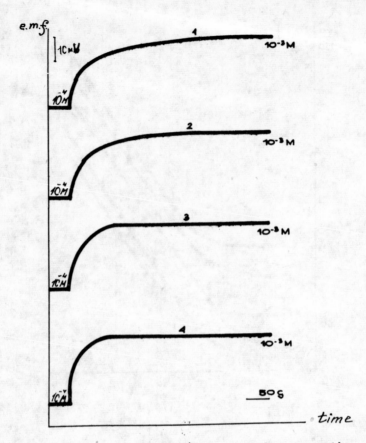

Fig. 5   Effect of the membrane organic solvent on the
          response times. 1- 1,2 dichlorethane; 2- chloroform;
          3- chlorex; 4- chlorbenzol

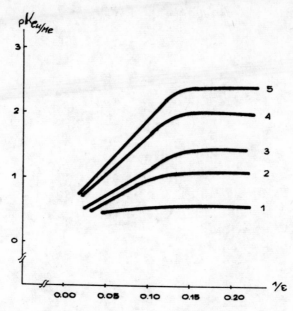

Fig. 6   Effect of the dielectric constants on the potentiometric selectivity coefficients $pK_{Cu/Me}$ for various
interfering cations. 1- Pb/II/; 2- Cd/II/; 3- Zn/II/;
4- Mn/II/; 5- Co/II/

PANEL DISCUSSION

PANEL DISCUSSION OF SELECTED TOPICS

Chairman: Prof. W. SIMON

Moderator: Prof. E. PUNGOR

1. Problems of Standardization in ISE
   Measurements

A brief introduction on setting up the single ion activity scales at high concentrations was given by Dr.Bates before the actual discussion. A detailed presentation of this subject can be found in "Ion-Selective Electrodes", Ed. Pungor, E., Akadémiai Kiadó, Budapest 1978.

In connection with it a detailed review of this topic is given by A.Hulanicki:

Ion Buffers for Standardization of Ion-Selective Electrodes

Calibration of ion-selective electrodes in the range below 10^{-5}-10^{-6} mol/dm^3 with solutions obtained from salts or by dilution of more concentrated standard solutions is undesirable because mostly it leads to serious errors or is even completely impossible for following main reasons:
- preparation of very dilute solutions of accurately known concentration is neither simple nor accurate,
- easy losses or contaminations of dilute solutions may give serious, positive or negative errors in electrode response,
- processes at the electrode-solution interface /adsorption, interaction with interstitial ions, redox processes etc./ may influence the electrode potential in the low concentration range.

To avoid these difficulties the use of ion-buffers has been proposed /1-3/. They find occasionally applications in analytical procedures, but are useful for studying the electrode characteristics both for better knowledge of electrode behaviour and for application of electrodes in physicochemical studies of ionic equilibria.

The principal requirements for ion-buffers are:

- accurately known activity or concentration of free /hydrated/ ions of interest, evaluated from known equilibrium constants.
- well established correlation between concentration and activity of free /hydrated/ ions;
- sufficient ion-buffer capacity to eliminate the effect of dilution or changes of the concentration of the ion of interest, due to accidental losses or contamination as well as due to processes occuring at the electrode--solution interface;
- sufficient pH buffer capacity, because buffers based on ligands with weakly basic properties are significantly pH sensitive.

The principle of buffering action is the formation of a slightly dissociated compound of the ion in question, and this may be achieved through a relatively stable soluble complex or through a slightly soluble precipitate in equilibrium with a common ion.

In the case of metal buffers the equilibrium concentration of metal ion is readily calculated on the basis of relevant equilibrium constants. If the concentration constant is presented as the conditional constant [4].

$$\beta'_n = \frac{[ML'_n]}{[M'][L']^n} = \beta_n \frac{\alpha_{ML_n}}{\alpha_M \alpha_L^n} \qquad /1/$$

where $\alpha_M = \left([M] + [MOH] + [M(OH_2)] + \ldots\right)\big/[M]$ /2/

$\alpha_L = \left([L] + [HL] + [H_2L] + \ldots\right)\big/[L]$ /3/

$\alpha_{ML_n} = \left([ML_n] + [MHL_n] + [M(OH)L_n] + \ldots\right)\big/[ML]$ /4/

414

The above expression is valid when the formation of poly-nuclear species is neglected, which is usually the case in dilute solutions, especially in presence of excess of ligand. The α -coefficients are calculated from levant equilibrium constants or taken from tables published elsewhere.

Assuming that the ligand is present in sufficient excess we can introduce following approximate relations:

$$\left[ML'_n\right] = c_M \; ; \quad \left[M'\right] = \alpha_M\left[M\right] \; ; \quad \left[L'\right] = c_L - nc_M$$

Introducing them, and rearranging the expression for the conditional constant we obtain

$$[M] = \frac{c_M}{\alpha_M \, \beta'_n \left(c_L - nc_M\right)^n} \qquad \qquad /5/$$

or in the logarithmic form
$$pM = \log\beta'_n + \log\alpha_M + n\log\left(c_L - nc_M\right) - \log c_M. \qquad /6/$$

In practical application of metal buffers it is advantageous to use systems where $n=1$, because of avoiding often incomplete stepwise complexation and because the effect of dilution which disappears when n equals unity. The maximal buffer capcity is attained when $c_n = \left(c_L - nc_M\right)$.

Eq. /6/ may be presented in a modified form using the right side of eq. /1/:
$$pM = \log\beta_n + \log\alpha_{ML} - n\log\alpha_L + n\log\left(c_L - nc_M\right) - \log c_M \qquad /7/$$
In this equation the terms $\log\alpha_{ML}$ and $n\log\alpha_L$ include the effect of pH. The former often equals zero, except strongly acidic or basic conditions, therefore protonation of the ligand, being a weak base, affects the value of pM of the metal buffer solution. This effect depends on the number of protons attached to the ligand in the region of application of the buffer. According to eq. /3/ to which the relevant constants have been introduced

$$\alpha_L = 1 + K_{H1}\left[H^+\right] + K_{H1}K_{H2}\left[H^+\right]^2 + \ldots \qquad /8/$$

the higher is the ligand protonated the more depends pM on pH.

28*

415

This indicates the necessity of simultaneous pH buffering, which can be achieved in several ways:
- when the pH of the metal buffer is close to pK value of the ligand, its excess works as a self buffering system;
- when another metal with weaker complexation properties is added in excess, the change of H^+ concentration is compensated through the second metal-ligand system, but the effect of dilution becomes more serious. Besides this system can be used only in the case if the electrode is immune to the second ion;
- when a separate pH buffer is added, but it must be taken into consideration that most pH buffer systems exhibit complexing properties towards metal ions. This can be neglected when the complexing ability of the pH buffer is at least by two orders of magnitude smaller than that of the metal-buffer.

The systems with a slightly soluble precipitate are not real buffers in the meaning as before. They should be treated rather as solutions having a known, often very small concentration of the ion in question. From the solubility product principle can be deduced similarly as before that:

$$pM = pK_{so} - n\log\alpha_X + n\log\left(C_X - nC_M\right) \qquad /9/$$

where X is the counter ion, forming a slightly soluble precipitate MX_n. From eq. /9/ follows that pH affects pM only when X is a weak base, and that dilution causes proportional, with the factor n, change of pM. The accidental change of C_X may be easily controlled when the concentration C_X is relatively large.

The reliability of pM buffer depends critically on several factors as:
- proper choice of equilibrium constants adjusted to needed conditions,
- lack of interfering solution reactions with other components as ionic strength adjustor, pH buffers etc.,
- lack of specific interactions between the membrane and buffer components. This should be checked for each type of membrane composition and generally cannot be predicted a priori [5],

- setting the proper concentration range, which not only depends on the buffer solution equilibria but also on the membrane. One of the procedures used for checking the validity of calibration with buffers is the coincidence of the calibration curve obtained by dilution of standard solutions with that obtained in ion-buffer solutions.

As examples of ion-buffers the following systems can be given:

for calcium [6,7]-NTA, EDTA

for copper [2,8], lead [9] and cadmium [10]-NTA, EDTA

for fluoride [3,11]-Th^{4+}, Zr^{4+}.

REFERENCES

1. D.D.Perrin, B.Dempsey, Buffers for pH and Metal Ion Control, Chapman and Hall, London, 1974

2. R.Blum, H.M.Fog, J.Electroanal.Chem., 34, 485 /1972/

3. E.Baumann, Anal.Chim.Acta, 54, 189 /1971/

4. A.Ringbom, Complexation in Analytical Chemistry, Interscience, New York, 1963

5. W.E. Van der Linden, G.J.M.Heijne, Proc.Conf. on ISE, Budapest 1977 /ed. E.Pungor/ Akadémiai Kiadó, Budapest 1978, p. 445

6. J.Růžička, E.H.Hansen, J.C.Tjell, Anal.Chim.Acta, 67, 155 /1973/

7. H.M.Brown, J.P.Pemberton, J.D.Owen, Anal.Chim.Acta 85, 261 /1976/

8. E.H.Hansen, C.G.Lamm, J.Růžička, Anal.Chim.Acta 59, 403/1973/

9. E.H.Hansen, J.Růžička, Anal.Chim.Acta, 72, 365 /1974/

10. J.Růžička, E.H.Hansen, Anal.Chim.Acta 63, 115 /1973/

11. M.Trojanowicz, Talanta 26, 985 /1979/.

E.Pungor:

At high concentrations we have to consider chemical reactions besides hydration, e.g. formation of polyanions.

R.G.Bates:

The standards have to be unassociated electrolytes, but the samples may contain polyanions.

E.Pungor:

We have also another problem with the calibration of ion--selective electrodes: the problem of ion-buffers. I wonder if we can set up a pMetal ion scale as useful as the pH scale, especially at low total concentrations of the determinand ion.

A.Hulanicki:

If the total concentration of the ion is very low, such buffers are not useful, either. But if we investigate some complex systems then we have to standardize the electrode in a known buffer system.

J.D.R.Thomas:

But with regard to copper in water there is also the problem of the measuring device, and I think that is probably a bigger problem in that case than buffer capacity.

W.Simon:

I want to stress the importance of having metal buffers for calibration, because we have a paradox situation. Whenever we calibrate the pH scale we take buffers. But with ISE's some people apparently have the feeling that no buffers are needed and this is just not possible, at least not at the lower concentration levels because these solutions would not be stable at all. For the use with ISE-s we have to prepare quite a wide selection of buffer solutions of different ionic strengths for special applications.

E.Pungor:

The calibration of ISE-s at low total concentrations is like
measuring pH between 6 and 8 in unbuffered solutions. Of course
we can obtain good results at low concentrations if a suitable
ionic buffer is available, but we cannot measure the activity
in the sample if the total concentration of the determinand is
very low. This fact is independent from the way of calibration.

G.Werner:

Buffer systems are also useful for the solution of analytical
problems where you intend to determine a metal in a buffered
solution. Only ISE-s can be used to do such investigation
without disturbing the equilibrium. The concentration of the
free metal ion can be determined with ISE-s. The problem is
to make the calibration buffer as similar as possible to the
investigated system.

R.G.Bates:

There are two separate problems here: the use of metal buffers
for calibrating the scale is one of them, the other one is
the difference between concentration and activity, but this
applies, of course, only at the high concentrations. One thing
that does bother me though, is when you use a metal buffer,
presumably the ionic strength is going to be fairly low,
say 0.1, but if you calculate the metal ion concentration
from the stability constant, you are estimating only a concentration and your electrode really responds to an activity.

E.Pungor:

It will be also important in the future to define an activity
scale, because we are measuring activities.

E.Juhász:

The problem here is similar to temperature measurement above
the gold point. There we use optical pyrometers to measure
emissivity by using theoretical equations which refer to
black body radiation. We are doing this even though the black
body cannot be perfectly realized. With ISE-s I would be con-

tent to have a well defined convention and some solutions the
activity value of which is measured on the basis of this
convention.

A.Lewenstam:
The situation with ISE-s is not as clear as with the temperature measurement. There is not only the problem of activities
in the solution but also the activity of each component in
the membrane should be established. We may have different
results in the some solution but with different electrodes.

2. Classification of ISE-s

R.P.Buck:
There is already an IUPAC convention on the naming of ISE-s.
We may consider also a classification of electrodes based on
their working mechanism. With some electrodes we have electron
transfer, with others ion transfer across the electrode/
electrolyte interface. If we regard as ISE-s only those
electrodes with ion transfer, a great part of potentiometric
electrodes would not fall into the category of ISE-s. I
prefer a classification which allows for all potentiometric
electrodes where there is a selectivity involved to be regarded
as ISE-s. We could say that any material which senses ion
activities is an ISE.

W.Simon:
I agree with Dr.Buck and I disagree with a classification based
on the mechanism of transport, because we would then have to
reclassify the electrodes after some time as the research goes
on.

E.Pungor:
Although we have difficult existing ways of classification
of ISE-s I think this is not a great problem because such a
classification is mainly needed to guide the beginners in the
field.

J.Koryta:

The classification of ISE-s is also needed in the field for
researchers. It is necessary to distinguish between the
classical types of electrodes /first kind, etc./ and ISE-s.
It is unreasonable to speak e.g. about the selectivity of a
Zn/Zn^{2+} electrode.

W.Simon:

The term ion-selective electrode should in my opinion include
electrodes like the Zn/Zn^{2+} electrode, while the expression
ion-selective membrane electrodes should be used for electrodes
with exclusive ion transfer.

E.Pungor:

Classification is first of all necessary for teaching electrochemistry. A possible classification can be as follows:
I.1. Redox electrodes /first order, etc./
  2. Ion-exchange electrodes /acid-base exchange,precipitate
     exchange, complex formation electrodes/
II. Sensitized electrodes /with active or inactive coverage/
This classification gives a clear picture for the purpose of
education.

R.P.Buck:

Is the present categorisation of the membrane types set for
ever by IUPAC or is it to be revised from time to time or it
will be kept as it is and we will to have to work around it ?

J.D.R.Thomas:

There may be a revision of the present system later on.

E.Pungor:

We should suggest to the IUPAC working party to revise the
classification of ISE-s.

J.Koryta:

A consultation with the electrochemists in IUPAC on the questions of nomenclature would be advantageous.

A.Lewenstam:

One may obtain a three dimensional classification of ISE-s if one considers three important characteristics of them at the some time, such a classification is complicated, however.

Interface of two immiscible
electrolyte solutions /ITIES/
simple ion transfer, in 54

LaF$_3$ electrodes 287
Lead-selective electrodes
    264, 265
Liquid ion-exchangers 35
- junction potentials 206,209
- membrane electrodes 73, 306,
    310
- - - memory effect of 361
- membranes, investigation
    of 399
- - with charged carrier 327
- - selectivity interferen-
    ces in 105, 107

Membrane cells, superficial
    structure of 31
Membranes of AgX-Ag$_2$S type
    150, 154, 156, 168
- of thin dielectric films
    150, 160, 173
Memory effect of ISE-s 359
Metal buffers for standar-
    dization of ISE-s 414
Metrological basis for
    calibration of ISE-s 203
Micro ion-selective elect-
    rodes 129
- immunoelectrode 349
Mixtures of electrolytes 8
Model calculation on the
    structure/selectivity 315,
    321, 322
Multiple ISE-detector 375

Natural gas, ethanethiol con-
    tent of 233
Neutral carrier membranes 73
- - - structure of 75, 80
- - - transport numbers of
    74, 79
- - sensors 125
Nicotinamide, determination
    of 300
Nikolsky equation 103
Nitrate ISE-s 125
Non-specific interferences
    104, 109

pH-sensitive ISFET-s 387
pK, pNa, pCl values at differ-
    ent temperatures 205, 207,
    208
Phospholipid monolayer at
    ITIES 58, 67-69
- liposomes 349
Photoelectron countrates 90,
    95
- , emission of 89, 93
- spectrum of lunar rock
    sample 90, 94
Polycrystalline AgX, Ag$_2$S 150,
    154, 155, 167
Potassium in urine, continuous
    monitoring of 179
- chloride, determination of
    300, 302
Potentiometric analysis, equi-
    librium reactions in 297
Proton transfer across ITIES
    57, 66
PVC calcium electrode 252,255